Undergraduate Lecture Notes in Physics

For further volumes:
www.springer.com/series/8917

Undergraduate Lecture Notes in Physics (ULNP) publishes authoritative texts covering topics throughout pure and applied physics. Each title in the series is suitable as a basis for undergraduate instruction, typically containing practice problems, worked examples, chapter summaries, and suggestions for further reading.

ULNP titles must provide at least one of the following:

- An exceptionally clear and concise treatment of a standard undergraduate subject.
- A solid undergraduate-level introduction to a graduate, advanced, or non-standard subject.
- A novel perspective or an unusual approach to teaching a subject.

ULNP especially encourages new, original, and idiosyncratic approaches to physics teaching at the undergraduate level.

The purpose of ULNP is to provide intriguing, absorbing books that will continue to be the reader's preferred reference throughout their academic career.

Sylvie Braibant · Giorgio Giacomelli ·
Maurizio Spurio

Particles and Fundamental Interactions: Supplements, Problems and Solutions

A Deeper Insight into Particle Physics

Springer

Sylvie Braibant
Department of Physics and INFN
University of Bologna
Bologna, Italy

Giorgio Giacomelli
Department of Physics and INFN
University of Bologna
Bologna, Italy

Maurizio Spurio
Department of Physics and INFN
University of Bologna
Bologna, Italy

ISSN 2192-4791 ISSN 2192-4805 (electronic)
Undergraduate Lecture Notes in Physics
ISBN 978-94-007-4134-8 ISBN 978-94-007-4135-5 (eBook)
DOI 10.1007/978-94-007-4135-5
Springer Dordrecht Heidelberg New York London

Library of Congress Control Number: 2012936363

Printed on acid-free paper

Springer is part of Springer Science+Business Media (www.springer.com)

Preface

The *Problems* presented here refer to the topics discussed in the corresponding 14 chapters of the *Particles and Fundamental Interactions* textbook. This *Problems, Solutions and Supplements* book is aimed at students in a course of Experimental Particle Physics, not only as a preparation for a written examination, but also as a necessary instrument for a deeper understanding of high energy physics. It contains 170 problems of different difficulty levels. Some of them are traditional, covering most aspects of particle properties and of their fundamental interactions, and some are more advanced. Some problems are derived from our teaching experience to undergraduate students; some are derived from the admission examination to the PhD courses in Italian universities; some are completely original, from our research activities.

Each problem has an identification number and a *title* to facilitate the identification of the subject discussed in the text. Most problems are solved step-by-step, to help both students and teachers to get better acquainted to topics presented in the textbook. We follow the same chapter numeration of the textbook. To avoid confusion when we refer to chapters, equations, figures and tables of *Particles and Fundamental Interactions*, the reference is enclosed in a box. In this way, Fig. 7.2 refers to a figure in this manual, and [Fig. 7.2] to a figure in the textbook. As a general advice, it is useful to try to solve problems only after a first reading of a book. Before facing the more advanced problems, we suggest to read at least up to Chap. 8 of *Particles and Fundamental Interactions*, where the introduction of particle nomenclature and classification, and the presentation of fundamental aspects of the interactions are completed.

In addition to problems and solutions, additional material is presented in form of fifteen *Supplements*. Four of them present the most powerful accelerators, those which produce *cosmic rays*. Cosmic rays were of fundamental importance for the discovery of most long-living particles, the development of particle physics and that of *astroparticle physics*. Three Supplements are devoted to the electronic signals, to data acquisition systems, to the electronic logics and triggers of the experimental apparatuses, ending with the computing effort required for the LHC collider. These issues play also a key role in the contribution that particle physics research provides as a spin-off technology. This is also true for other four supplements, presenting additional information on interactions between charged particles and matter (multiple Coulomb scattering, synchrotron radiation) or the use of radioactive decays for dating old objects. Some problems contain, after the solutions, some comments related to past, running or future experiments (as for instance that for the neutrino beams and neutrino oscillations, the search for proton decay, the study of symmetry violation through the electric dipole moment of the neutron, the measurement of α_S, the study of astrophysical objects using charged particles and/or neutrinos, etc.).

We thank many colleagues, in particular those of the former OPAL and MACRO groups (now, CMS, OPERA and ANTARES) at the University of Bologna, for their cooperation. Finally, we are grateful to many students for their suggestions and questions that allowed us to prepare this work in a way that we hope will be useful for many.

We are responsible for the errors which inevitably could be present in this manual. Some problems contain approximations, or may be solved in different or more straightforward way. We apologize in advance for any mistake that could have survived and that the readers will discover: you are kindly encouraged to inform us.

Bologna, Italy

Sylvie Braibant
Giorgio Giacomelli
Maurizio Spurio

Contents

Chapter 1
Historical Notes and Fundamental Concepts

Problems

1.1. **Orders of magnitude**. It is important to become familiar with the orders of magnitude typical of submicroscopic systems. The molecules have dimensions of the order of 10^{-7} cm, the hydrogen atom of the order of 10^{-8} cm, the proton and the neutron of the order of 10^{-13} cm. The size of a quark is less than a hundredth of the proton size. Although the proton can be considered as an almost empty system, it is worth recalling that the vacuum is a complex system.

(a) Let us hypothesize to be able to align the H_2O molecules of a cubic centimeter of water; what would be the length of the very thin line obtained?

(b) The atoms have a size of the order of 10^{-8} cm; an atom is essentially an empty system. If we imagine that the dimension of the proton is 1 mm, at what distance would be the electron in a hydrogen atom?

[A: (a) 20 times the distance Earth-Sun. (b) about 100 m]

1.2. **Natural units**. In the natural unit system ($\hbar = c = 1$), derive the dimensional relations between mass and length and between mass and time.
[A: $[M] = [L^{-1}] = [T^{-1}]$, see Appendix A2].

1.3. **Consequences of $k = 1$**. The Boltzmann constant is $k = 1.38066 \cdot 10^{-23}\ \mathrm{J\,K^{-1}}$ $= 8.6173 \cdot 10^{-14}\ \mathrm{GeV\ K^{-1}}$. Assuming that the Boltzmann constant is $k = 1$ and dimensionless the temperature has the dimensions of energy. Determine the temperature corresponding to 1 eV and to 1 GeV.
[1 eV $= 1.1605 \cdot 10^4$ K; 1 GeV $= 1.1605 \cdot 10^{13}$ K]

1.4. **Planck mass**. The (dimensionless) gravitation constant can be written as:

$$\alpha_G = \frac{G_N M^2}{\hbar c}$$

S. Braibant et al., *Particles and Fundamental Interactions: Supplements, Problems and Solutions*, Undergraduate Lecture Notes in Physics,
DOI 10.1007/978-94-007-4135-5_1, © Springer Science+Business Media Dordrecht 2012

(different authors insert a 4π factor).
(a) Verify that α_G is dimensionless.
(b) Evaluate α_G for the proton mass $M = m_p$.
(c) Evaluate the *Planck mass* (or Planck energy), M_{Pl}. The Planck mass is, by definition, the mass of the hypothetical particle that would produce $\alpha_G = 1$.
[See solutions]

1.5. **Planck length**. Starting from M_{Pl}, determine a quantity that has the dimension of a length $[L]$. This quantity is defined as the *Planck length*. The Planck mass fixes the energy scale where the unification of the gravitational interaction with the other three interactions (strong, electromagnetic and weak) should occur. In string theory, the Planck length is the natural scale for the string size.
[See solutions]

1.6. **Cross-section $e^+e^- \to \mu^+\mu^-$**. The cross-section for the $e^+e^- \to \mu^+\mu^-$ process is:

$$\sigma = \frac{4\pi}{3}\frac{\alpha_{EM}^2}{s} \quad (\text{in GeV}^{-2} \text{ for } \hbar = c = 1).$$

α_{EM} is the dimensionless electromagnetic coupling constant (called the *fine structure constant* in atomic physics) and $s = E_{cm}^2$. Express the cross-section in the c.g.s system.
[See solutions]

1.7. **Plank units**. The Planck units are units of measurement defined exclusively in terms of five universal physical constants: the Gravitational constant, G_N; the Planck constant, $\hbar$; the speed of light in vacuum, c; the Coulomb constant, $\frac{1}{4\pi\varepsilon_0}$; and the Boltzmann constant, k (see Appendix 5). Using the dimensional analysis, find the five so-called *base Planck units* of mass, length, time, electric charge and temperature.
[See solutions]

1.8. **Kinetic energy**. Evaluate the kinetic energy in TeV of a 10 mg mosquito, moving with a speed of 10 cm/s.
[A: 0.3 TeV]

1.9. **Lifetime and path length**. Instead of the particle lifetime τ_0, the distance, $d = c\tau_0$, travelled by the particle during its lifetime is sometime specified. Find the momentum of the particle for which this relation is true.
[See solutions]

1.10. **Energy = mass.** At which value of β the kinetic energy is equal to the particle rest mass m_0?
[See solutions]

Supplement 1.1: Cosmic Rays and Astroparticle Physics

Before the advent of particles accelerators, the study and discovery of elementary particles, up to the 1950s, was performed using Cosmic Rays (CRs). *Astroparticle physics* is the field of study of the universe using charged and neutral particles, in addition to the electromagnetic radiation.

One of the main questions in astroparticle physics is the origin of high-energy CRs. It was discovered at the beginning of last century that energetic charged particles (mainly high energy protons and heavier nuclei, see Fig. 1.1 [1H06]) produce showers of secondary particles when hitting the Earth atmosphere. While the energy spectrum of the cosmic rays can be measured up to very high energies, their origin remains unclear. There are many indications in favor of the galactic origin of the CR bulk (protons and other nuclei up to $\sim 10^{15} \div 10^{16}$ eV); however, it is not possible to directly correlate the directions of CRs impinging on Earth to astrophysical sources because CRs are deflected by galactic magnetic fields.

The energy spectrum shown in Fig. 1.1 spans from $\sim 10^9$ eV to more than 10^{20} eV and follows a broken power-law of the form:

$$\left[\frac{dN_P}{dE}\right]_{obs} = K \cdot E^{-\alpha} \quad (\mathrm{cm}^{-2}\,\mathrm{sr}^{-1}\,\mathrm{s}^{-1}\,\mathrm{GeV}^{-1}) \tag{1.1}$$

where $\alpha \simeq 2.7$ from $\sim 10^{10}$ eV up to $\sim 3 \times 10^{15}$ eV and $\alpha \simeq 3.1$ for $3 \times 10^{15} < E < 10^{19}$ eV. Up to energies of $\sim 10^{14}$ eV, the CR spectrum is directly measured above the atmosphere using stratospheric balloons or satellites. Measurements show that $\sim 90\%$ are protons, $\sim 9\%$ are Helium nuclei and $\sim 1\%$ are heavier nuclei. Electrons are about 1% of the protons. Below $10^2 \div 10^3$ TeV the mechanism responsible for the acceleration of particles is plausibly iterative scattering processes of charged particles in a shock-wave (the so-called Fermi model, which predicts the differential energy spectrum of accelerated particles). These shock-waves are originated in environments of exceptional disruptive galactic events, like stellar gravitational collapses (type II *supernovae*).

Accelerated charged particles are confined in the Galaxy by the galactic magnetic fields (of average value $B \sim 3\ \mu\mathrm{G}$) for a time of $\tau_D \sim (3 \div 10) \times 10^6$ years.[1] The gyromagnetic radius for a particle with charge Ze, energy E, in the magnetic field B is $R \simeq \frac{E}{eZB}$. For this reason, the confinement time τ_D is not constant but decreases as

---

[1]The time τ_D is also called *diffusion time* as it corresponds to the CR average time to escape from the Galaxy. The CR escape probability P_D is thus inversely proportional to the diffusion time, $P_D \sim 1/\tau_D$.

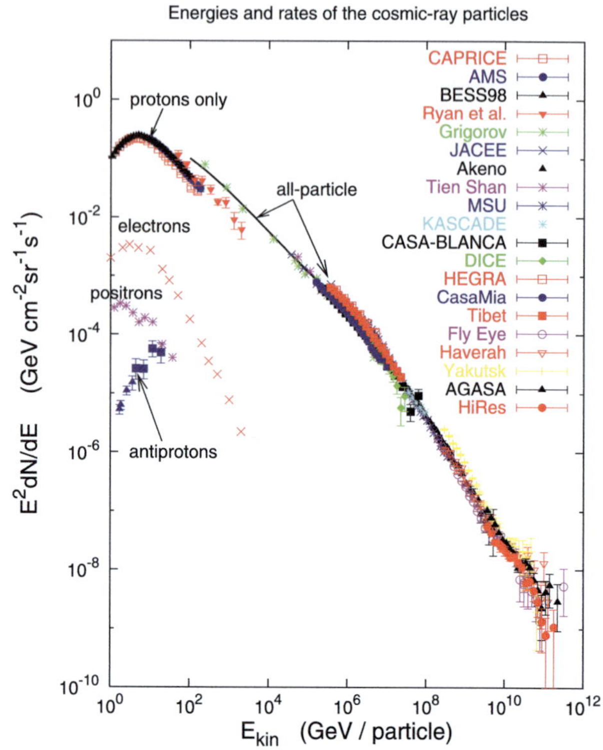

Fig. 1.1 Cosmic Ray spectrum from 10^9 to 10^{21} eV as measured on Earth. Note that the vertical scale is multiplied by E^2. In the low-energy region, when measurements are available, contributions of protons, electrons, positrons and antiprotons are reported separately [1H06]

the particle energy increases. During propagation, higher energy particles (at a fixed value of Ze) have a larger probability to escape from the Galaxy due to their larger gyromagnetic radii. It was found (see Supplement 14.1) that $\tau_D(E) \sim E^{-\alpha_D}$, with the diffusion exponent $\alpha_D \sim 0.6$ [1B95]. The measured spectral index ($\alpha \sim 2.7$) in Eq. (1.1) is steeper than the expected spectrum near the sources. The spectral index α_S of the energy spectrum at sources can be estimated using the measured spectrum (1.1) and the CR escape probability $P_D(E) \sim 1/\tau_D(E)$:

$$\left[\frac{dN_P}{dE}\right]_{sources} = \left[\frac{dN_P}{dE}\right]_{obs} \times P_D(E) \propto E^{-\alpha} \times E^{\alpha_D} \propto E^{-\alpha_S} \tag{1.2}$$

which gives $\alpha_S = \alpha - \alpha_D \simeq 2$, as predicted by the Fermi model [1F49a, 1F49b].

Above $\sim 10^{14}$ eV, CR measurements are only accessible from ground-based experiments [S04]. When CRs enter the Earth's atmosphere they collide with nucleons of atmospheric nuclei (mainly oxygen and nitrogen) and produce a cascade of secondary particles, the so-called *air shower*. The basic mechanism of air shower production of a CR proton on a nucleon N is the reaction:

$$p + N \rightarrow \pi^{\pm}, \pi^0, K^{\pm}, K^0, p, n, \ldots \text{(strange, exotic mesons and baryons)} \tag{1.3}$$

The decays of short-lived hadrons lead to a shower with three components [G90]:

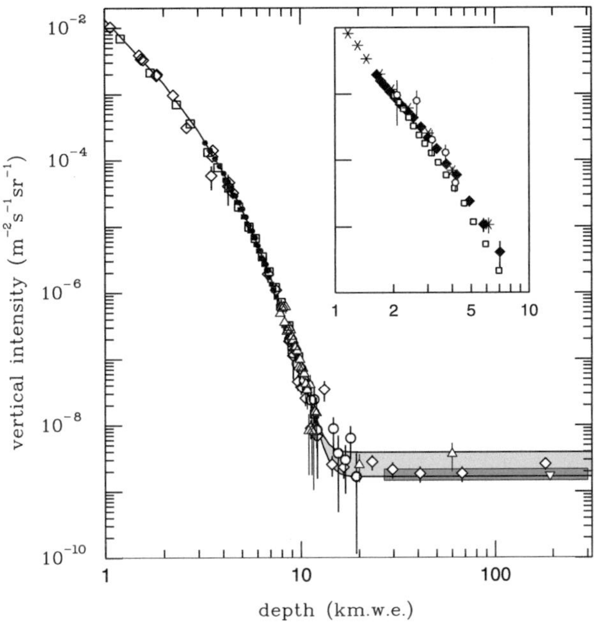

Fig. 1.2 Muon flux intensity (in $m^{-2}\ s^{-1}\ sr^{-1}$) from the vertical direction vs. depth (1 km.w.e. = $10^5\ g\,cm^{-2}$) of standard rock as measured by underground experiments. The *shaded area* at depths larger than ∼12 km.w.e. represents neutrino-induced muons of energy above 2 GeV. Only neutrinos can cross such amount of material [P10]

- the electromagnetic component of photons and electrons;
- the hadronic component of neutral and charged long-living hadrons;
- the penetrating component of muons and neutrinos.

Neutral meson decays (i.e., $\pi^0 \rightarrow \gamma\gamma$) lead to the development of the electromagnetic cascade; the γ's create electron-positron pairs which then emit bremsstrahlung photons in an avalanche process. The cascade stops when the typical particle energy becomes comparable with the critical energy of the medium (see Sect. 2.2.3). The penetrating component of *atmospheric muons* is detected also underground (or under many kilometers of water), see Fig. 1.2 [P10]. The *atmospheric neutrinos* are able to cross the whole Earth. The mechanism of *neutrino oscillations* (Chap. 12) was discovered using atmospheric neutrinos.

The showers of secondary particles created by interaction of primary CRs in the atmosphere are distributed in a large area. Different experimental techniques are applied to derive the flux of primary CRs through the measurement of secondary particles. The *Extensive Air-Showers Arrays* (EAS) are composed of a collection of detectors distributed on a large area. Scintillators or water-Cherenkov counters are typically used to detect the passage of charged particles reaching the ground. Other techniques include the *Cherenkov telescopes*, which detect the Cherenkov light emitted by the electrons in the atmosphere and the *fluorescence detectors*, which observe the fluorescence light emitted by atmospheric nitrogen excited by the shower particles. These detectors can estimate many characteristics of the shower: the number of secondary particles, related to the primary CR energy; the shower lateral distribution with respect to the axis; the primary CR direction of incidence.

The measured CR energy spectrum above $\sim 10^{19}$ eV gives a spectral index $\alpha \sim 2.5$. The flux, still dominated by protons and nuclei, is one particle per kilometer squared per year per stereoradian. It has long been assumed that these ultra high energy (UHE) CRs have an extragalactic origin, and can be detected only by very large ground-based installations [1K11].

The search for UHE CR sources must take into account another effect, the Greisen-Zatsepin-Kuzmin cutoff (GZK), which imposes a theoretical upper limit on the energy of cosmic rays from distant sources (see Problem 13.7). Nowadays, the largest CR array experiment is the Auger Observatory (located in Argentina), which combines the measurement of extensive air showers and light fluorescence detection. It covers a surface of ~ 3000 km^2.

Additional information on CR sources and propagation in our Galaxy can be found in Supplements 13.1 and 14.1.

Solutions

Problem 1.4

(a) Let analyze the dimensions of the numerator and denominator:

$$\begin{aligned}
[G_N] &= [Energy\ L M^{-2}] = [L^3 M^{-1} T^{-2}],\\
G_N &= 6.673 \cdot 10^{-11} \mathrm{m}^3\,\mathrm{kg}^{-1}\mathrm{s}^2\\
[G_N\, M^2] &= [L^3 M T^{-2}] = [M L^2 T^{-2} L] = [Energy \cdot L]\\
[\hbar c] &= [Energy \cdot L].
\end{aligned}$$

(b) In the International System (IS), $m_p = 1.6726 \cdot 10^{-27}$ kg. It follows that:

$$\alpha_G = \frac{G_N m_p^2}{\hbar c} = \frac{6.6726 \cdot 10^{-11} \cdot (1.6726 \cdot 10^{-27})^2}{1.0546 \cdot 10^{-34} \cdot 3 \cdot 10^8} \simeq 5.90 \cdot 10^{-39}.$$

(c) The Planck Mass is:

$$\begin{aligned}
M_{Pl} = \sqrt{\frac{\hbar\, c}{G_N}} &\simeq \sqrt{\frac{1.0546 \cdot 10^{-34} \cdot 3 \cdot 10^8}{6.673 \cdot 10^{-11}}}\\
&\simeq 2.177 \cdot 10^{-8}\ \mathrm{kg}/1.7827 \cdot 10^{-36}\ \mathrm{eV/c^2\,kg}\\
&\simeq 1.221 \cdot 10^{28}\ \mathrm{eV/c^2} \to 1.221 \cdot 10^{19}\ \mathrm{GeV/c^2}.
\end{aligned}$$

Problem 1.5 In the natural unit system, one has $[M] = [L^{-1}]$ (Problem 1.2). Therefore, it follows that:

$$\begin{aligned}
l_{Pl} = \frac{1}{M_{Pl}} \to \frac{\hbar}{M_{Pl}\, c} = \frac{\hbar\, c}{M_{Pl}\, c^2} &= \frac{0.19733\ \mathrm{GeV\ fm}}{1.221 \cdot 10^{19}\ \mathrm{GeV}}\\
&= 1.616 \cdot 10^{-20}\ \mathrm{fm} = 1.616 \cdot 10^{-35}\ \mathrm{m}.
\end{aligned}$$

In terms of G_N, one has: $l_{Pl} = \sqrt{\hbar G_N / c^3}$.

Recalling that $\lambda\!\!\!^{-}_e = \hbar / m_e c$ is the electron Compton wavelength, the Planck length corresponds to the Compton wavelength associated with a particle of mass M_{Pl}.

Problem 1.6 Since $[\sigma] = [L^2]$, the cross-section formula given in the problem must be multiplied by a factor containing $\hbar$ and c and with the dimensions of $[Energy^2]\,[L^2]$. This factor is $(\hbar c)^2$.

$$\sigma = \frac{4\pi}{3} \frac{\alpha^2_{EM}}{s} (\hbar c)^2 = \frac{4\pi}{3} \frac{1}{(137.04)^2} \frac{(0.19733)^2 \text{ GeV}^2 \text{ fm}^2}{s \qquad \text{GeV}^2}$$

$$\simeq \frac{86.8 \cdot 10^{-7}}{s} \text{ fm}^2 = \frac{86.8}{s} \cdot 10^{-7} \cdot 10^{-26} \text{ cm}^2 = \frac{86.8}{s} \text{ nb}.$$

σ is in nb if s is expressed in GeV^2. The same result is obtained using direct conversion factors:

$$1 \text{ GeV}^{-1} = 1.9733 \cdot 10^{-16} \text{ m} = 1.9733 \cdot 10^{-14} \text{ cm}$$

$$1 \text{ GeV}^{-2} = 3.894 \cdot 10^{-28} \text{ cm}^2 \simeq 3.894 \cdot 10^{-4} \text{ b}$$

$$\sigma = \frac{4\pi}{3} \frac{\alpha^2_{EM}}{s} \text{ GeV}^{-2} \simeq \frac{4\pi}{3} \frac{1}{(137.04)^2} \frac{3.894 \cdot 10^{-4} \text{ b}}{s} = \frac{86.8}{s} \text{ nb}.$$

Problem 1.7

Name	Dimension	Formula	Value (I.S.)
Planck length	$[L]$	$l_P = \sqrt{\frac{\hbar G_N}{c^3}}$	1.616252×10^{-35} m
Planck mass	$[M]$	$m_P = \sqrt{\frac{\hbar c}{G_N}}$	2.17644×10^{-8} kg
Planck time	$[T]$	$t_P = \frac{l_P}{c} = \sqrt{\frac{\hbar G_N}{c^5}}$	5.39124×10^{-44} s
Planck charge	$[Q]$	$q_P = \sqrt{4\pi\varepsilon_0 \hbar c}$	$1.875545870 \times 10^{-18}$ C
Planck temperature	$[T]$	$T_P = \frac{m_P c^2}{k_B} = \sqrt{\frac{\hbar c^5}{G_N k_B^2}}$	1.416785×10^{32} K

Problem 1.9 The distance travelled by a particle with speed $v = \beta c$ during the time τ_0 is $d = \beta c \gamma \tau_0$. Remember that $p = m \beta c \gamma$, then $\beta c \gamma = p/m$. It follows:

$$d = \beta c \gamma \tau_0 = p/m \tau_0.$$

The relation $d = c\tau_0$ is valid when $p/mc = 1$. For instance, for a muon, $d = c\tau_0 =$ 658 m is the average distance travelled by a muon of momentum $p = 105$ MeV/c.

Problem 1.10 The kinetic energy is defined as $T = E - m_0 = \gamma m_0 - m_0$ where $\gamma = \frac{1}{\sqrt{1-\beta^2}}$. The kinetic energy value is equal to the rest mass m_0 when

$$T = \gamma m_0 - m_0 = m_0.$$

This correspond to $\gamma m_0 = 2m_0$ or $\gamma = 2$. Therefore, one has:

$$\gamma = \frac{1}{\sqrt{1-\beta^2}} = 2 \longrightarrow \beta = 0.86.$$

References

[1F49a] Fermi, E.: On the origin of cosmic rays. Phys. Rev. **75**(1169), 15 (1949)
[1F49b] Fermi, E.: Galactic magnetic fields and the origin of cosmic radiation. Astrophys. J. **119**, 1–6 (1954)
[1B95] Biermann, P.L., Gaisser, T.K., Stanev, T.: Origin of galactic cosmic rays. Phys. Rev. D **51**, 3450 (1995)
[1H06] Hillas, A.M.: Recent progress and some current questions. astro-ph/0607109. The figure is due to T. Gaisser
[1K11] Kotera, K., Olinto, A.V.: The astrophysics of ultrahigh-energy cosmic rays. Annu. Rev. Astron. Astrophys. **49**, 119–153 (2011)

Chapter 2
Particle Interactions with Matter and Detectors

Problems

2.1. **Atom number density**. In the interaction of particles (or nuclei) with matter, the number of collisions depends on the number of scattering centers per unit volume. Often, the scattering centers are atomic nuclei. Consider for example the case of carbon, which has an atomic mass number $A = 12$ and a density (specific mass) $\rho \simeq 2.265\ \text{g}\,\text{cm}^{-3}$. Determine:
(a) the number of atoms per cm^3;
(b) the number of atoms per gram.
[See solutions]

2.2. α **particle energy loss**. An α particle with 7.4 MeV kinetic energy crosses a target consisting of a thin copper foil $5 \cdot 10^{-4}$ cm thick. Determine:
(a) the ionization energy loss in the copper foil;
(b) the particle kinetic energy and (c) the Coulomb multiple scattering angle when going out of the foil.
Hint: see Supplement 2.1.
[See solutions]

2.3. **Muon Energy loss**. A muon of 100 GeV energy crosses without being absorbed a detector whose mass is mainly due to the hadronic calorimeter and to the muon detector. The thickness of the crossed material can be considered as a layer of 3 m of iron. Determine:
(a) what is the dominant energy loss process;
(b) the average energy loss of the muon inside the detector.
Hint: see Supplement 2.2.
[See solutions]

S. Braibant et al., *Particles and Fundamental Interactions: Supplements, Problems and Solutions*, Undergraduate Lecture Notes in Physics,
DOI 10.1007/978-94-007-4135-5_2,

2.4. **Energy transferred**. Calculate the maximum energy ν_{max} transferred in elastic scattering of a charged particle with mass M and energy $E = T + Mc^2$ to an electron at rest:
(a) in the non relativistic case ($T \ll Mc^2$);
(b) in the relativistic case with $M \gg m_e$;
(c) in the general case.
[A: (c) $\nu_{max} = \frac{2m_e c^2(E^2 - M^2c^4)}{M^2c^4 + m_e^2c^4 + 2Em_ec^2}$]

2.5. **Kinematics of the Compton effect**. Using the energy and momentum conservation, describe the kinematics of the Compton effect and derive Eq. (2.19).
Calculate the maximum energy of the recoiling electron Eq. (2.21).

2.6. **Electromagnetic shower**. Calculate the average number of particles in an electromagnetic shower initiated by a 50 GeV photon, after 10, 13 and 20 cm of crossed iron.
[See solutions]

2.7. **Muon from pion decay**. Consider a π^+ at rest decaying in $\pi^+ \to \mu^+\nu_\mu$. Calculate the μ^+ kinetic energy and evaluate approximately the μ^+ range in liquid hydrogen (specific mass $\rho = 0.07$ g cm^{-3}).
[See solutions]

2.8. **Neutron discovery**. In his Letter to the Editor of Nature of February 27, 1932 (*Possible Existence of a Neutron*), J. Chadwick described the observation of protons emitted from a target containing hydrogen atoms. The hydrogenated target was exposed to an unknown radiation of strong penetrating power emitted by beryllium when bombarded by α-particles from polonium. See the layout presented in Fig. 2.1. The protons (with mass m_p) were emitted with velocities up to a maximum of nearly 3×10^9 cm/s. Since the penetrating radiation emitted by the beryllium was observed to be neutral, it could consist either of photons or, according to Chadwick's hypothesis, of neutral particles with a mass similar to that of the proton, i.e., the neutrons. Assuming that the neutral radiation emitted by the Be is composed of photons and that the protons are emitted through the Compton effect induced by these incident photons, calculate the photon energy E_γ. Discuss why this E_γ is inconsistent with the observation. Finally, discuss the reasons that led Chadwick to formulate the hypothesis of the neutron existence.
[See solutions]

2.9. **Multiple Scattering-1**. Calculate the Coulomb multiple scattering angle in the plane θ^0_{plane} for protons
(a) of 50 MeV/c momentum in 0.1 g cm^{-2} of aluminum;
(b) of 200 MeV kinetic energy in 2 mm of copper.
[See solutions]

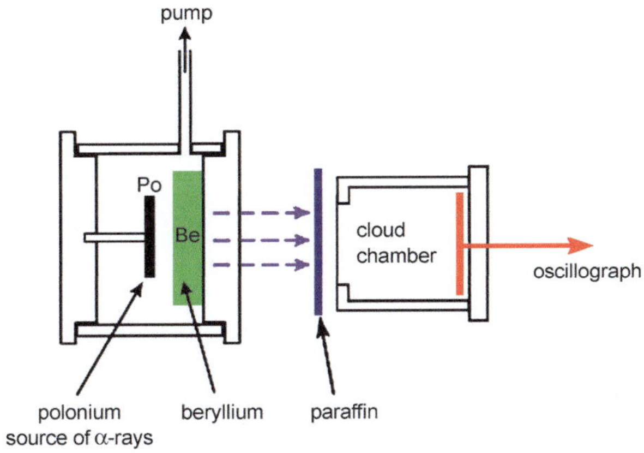

Fig. 2.1 Layout of Chadwick experimental apparatus that led to the neutron discovery. A beryllium target is exposed to high-energy α rays from a polonium source. A strong penetrating power radiation is emitted from the Be and hits the protons contained in the paraffin layer. The emitted protons are observed in the cloud chamber on the *right* [2w3]

2.10. **Multiple Scattering-2**. From considerations based on the Coulomb multiple scattering on nuclei, determine when a target is thin or thick.
[See solutions]

2.11. **Neutron moderation**. Neutrons produced in nuclear reactors are emitted with energies of order of a few MeV and must be slowed down to thermal energies through elastic scattering on nuclei of a *moderator*. Determine the neutron speed variation in each collision assuming that the moderator is (a) hydrogen; (b) carbon; (c) iron. Show that a non-relativistic calculation is sufficient.
[See solutions]

Supplement 2.1: Multiple Scattering at Small Angles

A charged particle traversing a medium is deflected by many small-angle scatters. This deflection is due to the superposition of many Coulomb scattering from individual nuclei, and hence the effect is called *multiple Coulomb scattering*. When the particle is a hadron, the strong interaction also contributes. The cumulative effect (for thick targets) is a deflection as that shown in Fig. 2.2.

For small deflection angles, the Coulomb scattering distribution is well represented by a Gaussian distribution. At larger angles (i.e., larger than the angle θ_0 defined below), the distribution shows larger tails and the behavior is more similar to that of the Rutherford scattering. In many applications, scattering at large angles is negligible and the Gaussian approximation for small angles describes well enough the projected angle distribution, with a width [P10]:

$$\theta_0 = \theta_{plane}^{rms} = \frac{13.6\,\mathrm{MeV}}{\beta c p} z \sqrt{\frac{x}{X_0}} \Big[1 + 0.038 \ln(x/X_0) \Big] \tag{2.1}$$

where p, βc, and z are respectively the momentum, velocity, and charge number of the incident particle; x/X_0 is the thickness of the scattering medium in units of

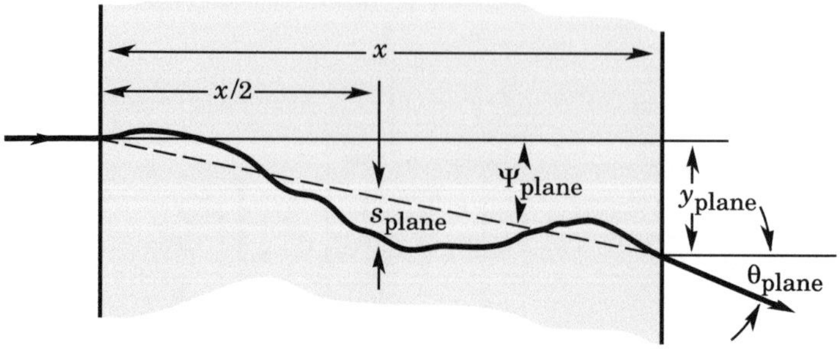

Fig. 2.2 Quantities used to describe the multiple Coulomb scattering. The particle is incident from the *left* in the plane of the figure. From [P10]

radiation length (Eq. (2.15)). The distribution of the angle in the space has width $\theta^{rms}_{space} = \sqrt{2}\theta^{rms}_{plane}$.

Supplement 2.2: Muon Energy Loss at High Energies

As for electrons (see Sect. 2.2.3), at sufficiently high energies, radiative processes become more important than ionization for all charged particles. In particular for muons, the critical energy occurs at several hundred GeV [2G01]. Radiative effects dominate the energy loss of energetic muons found in cosmic rays or produced at high energy accelerators. Radiative effects are characterized by small cross-sections, hard spectra, large energy fluctuations, and generation of electromagnetic or (in the case of photonuclear interactions) hadronic showers. Above the critical energy, the treatment of energy loss as a uniform and continuous process is inadequate. It is convenient to write the average muon energy loss rate as:

$$-dE/dx = a(E) + b(E)E \tag{2.2}$$

where $a(E)$ is the ionization energy loss, and $b(E)$ is the sum of energy losses due to e^+e^- pair production, bremsstrahlung, and photonuclear processes. In most approximations, the quantities a, b can be considered constant and independent of the muon energy E. In this case, the mean range x_0 of a muon with initial energy E is obtained by integrating Eq. (2.2):

$$\int_0^{x_0} dx = \int_E^0 dE/(a+bE) \longrightarrow x_0 \simeq (1/b)\ln(1+E/E_c^\mu) \tag{2.3}$$

where $E_c^\mu = a/b$. $b(E)$ can be computed for different materials; it changes only very slowly with energy. In water, b ranges between $(2\div4)\cdot 10^{-6}$ g^{-1} cm^2 for muon energies between $10^2\div10^7$ GeV. In standard rock, b is 20%÷30% higher than in water. Since $a(E) \sim 2$ MeV g^{-1} cm^2, the critical energy $E_c^\mu = a/b \sim 500$ GeV and the radiative losses dominate above several hundred GeV. The rates of energy loss for positive muons in copper as a function of $\beta\gamma = p/Mc$ over nine orders of magnitude in momentum is reported in Fig. 2.3.

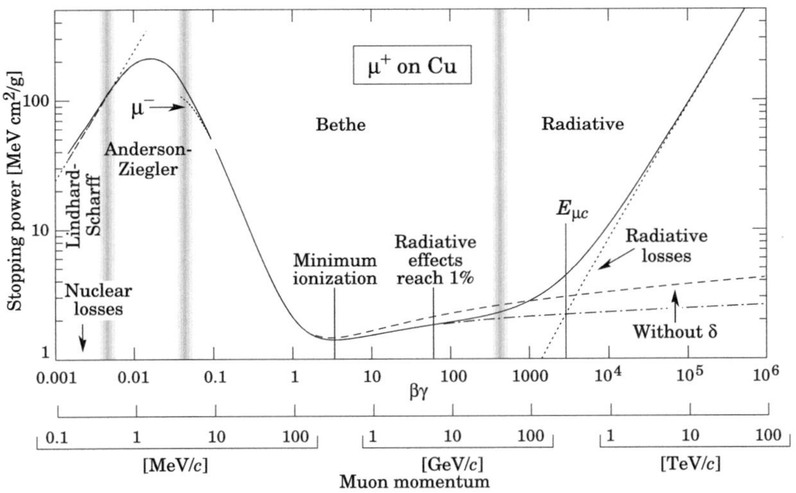

Fig. 2.3 Total energy loss $-(dE/dx)$ (*solid curves*) for positively charged muons in copper as a function of the muon momentum [P10]

Solutions

Problem 2.1 If the scattering centers are the material atoms (or atomic nuclei), one has:

$$\left(N_c = \frac{\text{scattering centers}}{\text{cm}^3}\right) = \left(N_a = \frac{\text{atoms}}{\text{cm}^3}\right) = \left[\frac{\text{n. of gram-molecule}}{\text{cm}^3} \cdot N_A \cdot H\right]$$

where H is the number of atoms per molecule, N_A is *Avogadro's number*, i.e., the number of molecules in a gram-molecule. The number of atoms per cm^3 is $N_a = \frac{m}{M}\frac{1}{\text{v}}N_A H = \frac{\rho}{M}N_A H$, where m is the mass in gram; M is the molecular weight in gram, v is the volume in cm^3 and $\rho = m/\text{v}$ is the specific mass. In the case of a monatomic element, for which $H = 1$ and $M = A$ ($A =$ atomic mass), one has $N_a = \rho N_A/A$. In the case of carbon for example, one has $A = 12$ and $\rho \simeq 2.265$ g cm^{-3}, and the number of atoms per cm^3 is:

$$N_a = \frac{\rho N_A}{A} \simeq \frac{2.265 \cdot 6.03 \cdot 10^{23}}{12} \frac{\text{g}}{\text{cm}^3} \frac{\text{molecule}}{\text{g moles}} = 1.137 \cdot 10^{23} \frac{\text{carbon atoms}}{\text{cm}^3}.$$

The number of atoms per gram is:

$$\frac{N_a}{\rho} \simeq \frac{1.137 \cdot 10^{23}}{2.265} \frac{\text{atoms}}{\text{g}} = 5.02 \cdot 10^{22} \frac{\text{carbon atoms}}{\text{g}}.$$

Problem 2.2 Consider the energy loss given in Eq. (2.9):

$$-\frac{dE}{dx} = \frac{4\pi z^2 e^4}{m_e v^2} N_e \ln \frac{\gamma^2 m_e v^3}{z e^2 \overline{\nu}}. \tag{2.4}$$

Taking into account that $N_e = N_A Z\rho/A$ and that $r_e = e^2/m_e c^2$, Eq. (2.4) can be rewritten as:

$$-\frac{dE}{dx} = 4\pi r_e^2 m_e c^2 \frac{N_A Z\rho}{A}\frac{z^2}{\beta^2}\ln\frac{m_e c^2\beta^2\gamma^2}{I} = C\rho\frac{Z}{A}\frac{z^2}{\beta^2}\ln\frac{m_e c^2\beta^2\gamma^2}{I} \tag{2.5}$$

where $C = 4\pi r_e^2 m_e c^2 N_A$ is a constant numerically equal to $C = 0.30$ MeV/g cm^{-2} and I is the average ionization potential that may be parameterized as $I = 13.6 \cdot Z$ eV (13.6 eV is the hydrogen ionization potential). In Eq. (2.5), the energy loss is factorized in three terms: the constant C, the term $(\rho\frac{Z}{A})$ which depends on the crossed material, the term $(\frac{z^2}{\beta^2})$ which depends on the particle charge and β times a logarithmic term which slightly depends on the particle $\beta\gamma$.

For copper, one has $\rho_{Cu} = 8.9$ g cm^{-3}, $Z = 29$, $A = 64$. For the considered α particle, one must determine the γ and β values from its known kinetic energy $T = E - m_\alpha$. For $c = 1$, one has (in natural units):

$$\beta\gamma = \frac{p}{m_\alpha} = \frac{\sqrt{E^2 - m_\alpha^2}}{m_\alpha} \simeq \sqrt{\frac{2T}{m_\alpha}} = \sqrt{\frac{2 \cdot 7.4}{3700}} = 0.064.$$

It is straightforward to verify that $\gamma = E/m_\alpha \simeq 1$. Placing these values in Eq. (2.5), one obtains:

$$\begin{aligned}-\frac{dE}{dx} &= C\rho\frac{Z}{A}\frac{z^2}{\beta^2}\ln\frac{m_e c^2\beta^2\gamma^2}{I} = 0.30 \cdot 8.9\frac{29}{64}\frac{2^2}{0.064^2}\ln\frac{0.511 \cdot 10^6 \cdot 0.064^2 \cdot 1^2}{13.6 \cdot 29} \\ &= \frac{4.84\ \text{MeV/cm}}{0.064^2}\ln(5.14) = 1997\ \text{MeV/cm}.\end{aligned} \tag{2.6}$$

When passing through a thickness of $5 \cdot 10^{-4}$ cm, the total energy loss is:

$$\Delta E = 1997\ \text{MeV/cm} \times 5 \cdot 10^{-4} = 1.0\ \text{MeV}.$$

(b) $T' = T - 1.0 = 6.4$ MeV

(c) Let us use Eq. (2.1) and take into account that the radiation length of copper (Table 2.1) corresponds to a path of 1.43 cm. The particle momentum is:

$$p = \sqrt{(T + m_\alpha)^2 - m_\alpha^2} \simeq \sqrt{2Tm_\alpha} = 234\ \text{MeV/c}.$$

Therefore (in the plane perpendicular to the motion), one has:

$$\theta_0 = \frac{13.6\ \text{MeV}}{0.064 \cdot 234\ \text{MeV}} \cdot 2 \cdot \sqrt{\frac{5 \cdot 10^{-4}}{1.43}}(1 - 0.30) = 23\ \text{mrad}$$

and $\theta_0^{space} = \sqrt{2}\theta_0 = 32$ mrad.

Problem 2.3

(a) As shown in Fig. 2.3, at the momentum of 100 GeV/c (remember that in the relativistic range $E = pc$) the dominant energy loss process is still that of excitation-ionization, with $dE/dx \sim 3$ MeV cm^2/g in the case of copper ($Z = 29$, $A = 64$). Since the energy loss depends only on the ratio Z/A of the crossed medium, it does not change for iron ($Z = 26$, $A = 56$).

(b) The density of iron is 7.87 g/cm^3, and then the muon energy loss in 3 m of iron is on average

$$\Delta E = 0.003\ [\text{GeV cm}^2/\text{g}] \cdot 7.87\ [\text{g/cm}^3] \cdot 300\ [\text{cm}] = 7.1\ \text{GeV}.$$

Problem 2.6 The processes considered here are the e^+e^- pair creation from photons, and the bremsstrahlung of electrons and positrons (i.e., the radiation of a high energy photon and the consequent energy decreases of the electron or positron). In both cases (see Fig. 4.7), the pair production and bremsstrahlung processes can be approximated as a process corresponding, on average, to the production of two particles sharing half the energy of the *parent particle*. The process stops when the particle energy drops below the *critical energy*. From that point on, the particles do not lose energy by pair production or bremsstrahlung (with increasing number of particles), but through the excitation and ionization processes.

The residual energy of a particle which has crossed a section of material of thickness x is given in Eq. (2.14). The radiation length and the path length of particles in iron (given in Table 2.1) are respectively 13.84 g cm^{-2} and 1.76 cm. After 10 cm of iron, the average energy of each particle is:

$$E_{10} = E_0 e^{-x/L_{rad}} = (5 \times 10^4) \cdot e^{-10/1.76} = 170.4\ \text{MeV}$$

higher than the critical energy $E_c = 27.4$ MeV in iron (see again Table 2.1). Since (on average) all particles have the same energy, the number of particles in the shower is:

$$n_{10} = E_0/E = \frac{5 \times 10^4}{170.4} = 293.$$

Applying the same calculation after 13 cm of iron, one finds $E_{13} = 31.0$ MeV. This value is slightly larger than the critical energy, and the corresponding number of particles is $n_{13} = 1613$. For an iron thickness larger than 13.2 cm, the average energy of the particles becomes smaller than E_c. The multiplicative process becomes less important with respect to the continuous energy loss mechanism. At a distance of 20 cm, the number of "surviving" particles is less than n_{13}.

Problem 2.7 The four-vector of the particles involved in the $\pi^+ \to \mu^+ \nu_\mu$ decay at rest are:

$$(m_\pi, 0) \to (E_\mu, \mathbf{p}_\mu) + (p_\nu, \mathbf{p}_\nu).$$

The neutrino mass is null (or completely negligible at this energy scale), and $E_\nu = p_\nu$. The condition for the momenta of the final state particles is simply:

$$|p_\nu| = |p_\mu|$$

while for the energy, one has:

$$m_\pi = E_\mu + E_\nu = E_\mu + |p_\nu| = E_\mu + |p_\mu| \longrightarrow |p_\mu| = m_\pi - E_\mu.$$

Finally, E_μ can be calculated using the mass-energy-momentum relation:

$$m_\mu^2 = E_\mu^2 - p_\mu^2 = E_\mu^2 - (m_\pi - E_\mu)^2$$

from which, one finds:

$$E_\mu = \frac{m_\mu^2 + m_\pi^2}{2m_\pi} = \frac{(105)^2 + (138)^2}{2 \times 138} = 109 \text{ MeV}.$$

The momentum of the emitted muon is:

$$p_\mu = m_\pi - E_\mu = 138 - 109 = 29 \text{ MeV/c}.$$

Figure 2.3 allows to determine the range of particles with known momentum. In fact, in this case, one has:

$$\beta\gamma = p_\mu / m_\mu = 29/105 \simeq 0.3$$

which corresponds to a value of $R/M = R/m_\mu = 1 \text{ g cm}^{-2} \text{ GeV}^{-1}$ (from inspection of the figure). Taking into account the specific mass of liquid hydrogen ($\rho = 0.07$ g cm^{-3}) and the muon mass ($m_\mu = 0.105$ GeV), one has:

$$\text{range} = \frac{R/m_\mu}{\rho} \cdot m_\mu = \frac{1 \times 0.105}{0.07} \simeq 1.5 \text{ cm}.$$

Problem 2.8 Protons emitted by the hydrogenated target have maximum velocity $\beta = v/c = 0.1$ and maximum momentum (in natural units, $c = 1$) $p_p = m_p\beta = 938 \times 0.1 \simeq 94$ MeV. The maximum kinetic energy T_p of the proton is:

$$E = T_p + m_p = \sqrt{m_p^2 + p_p^2} = \sqrt{938^2 + 94^2} = 942.7 \text{ MeV} \rightarrow T_p = 4.7 \text{ MeV}.$$

Let us assume that the protons in the hydrogenated target are extracted through Compton elastic scattering from photons coming from the beryllium. The photon energy $h\nu$ can be derived from the kinematics of the Compton effect given in Eq. (2.21). The maximum energy of the scattered particle is:

$$T_p = h\nu \frac{2\Gamma}{1 + 2\Gamma} \quad \text{with } \Gamma = \frac{h\nu}{m_p}.$$

Therefore, one can write:

$$T_p = \frac{2(h\nu)^2}{m_p + 2h\nu}.$$

Denoting $h\nu \equiv x$, this corresponds to a second degree equation:

$$2x^2 - 2xT_p - m_pT_p = 0 \rightarrow x = \frac{2T_p \pm \sqrt{4T^2 + 4 \cdot 2 \cdot m_pT_p}}{4}.$$

The solution with the negative sign must be excluded because it gives negative x. Thus, one has:

$$x \equiv h\nu = \frac{2T + \sqrt{4T^2 + 4 \cdot 2 \cdot m_pT_p}}{4} \simeq \frac{9.4 + \sqrt{8 \cdot 938 \cdot 4.7}}{4} = 51 \text{ MeV}.$$

Using energy and momentum conservation laws, Compton concluded that the radiation emitted by the beryllium is incompatible with the hypothesis of photons. The γ radiation emitted by exited nuclei is indeed below 10 MeV.

A more likely solution is that the neutral radiation incoming to the hydrogenated target is made of neutral particles with a mass similar to that of the proton. The elastic scattering between two particles with equal mass (one moving and one at rest) allows the transfer of the whole kinetic energy of the moving particle to the particle at rest (see Problem 2.11). Therefore, a *neutron* (the neutral counterpart of the proton) emitted from the Be target with a kinetic energy of $\sim$4.7 MeV is able to transfer such an energy to a proton at rest.

Problem 2.9

(a) The deflection of charged particles due to the multiple scattering is discussed in Supplement 2.1. According to Table 2.1, the radiation length of Aluminum is $X_0^{Al} = 24.0\ \mathrm{g\,cm^{-2}}$. Here, $x = 0.1\ \mathrm{g\,cm^{-2}}$. The width of the projected angle distribution is given in Eq. (2.1). To evaluate βpc, remember that $p = m\beta c\gamma$; $E = mc^2\gamma$ and thus $pc/E = \beta$. The energy for a $p = 50$ GeV/c proton is

$$E = \sqrt{p^2 + m_p^2c^4} \simeq m_pc^2 = 938\ \mathrm{MeV} \longrightarrow \beta = \frac{pc}{E} = \frac{50}{938} = 0.053.$$

This corresponds to $\beta pc = 0.053 \times 50\ \mathrm{MeV} = 2.66\ \mathrm{MeV}$.

The width of the projected angle is equal to:

$$\theta_0^a = \frac{13.6}{2.66}\sqrt{\frac{0.1}{24.0}}(1 - 0.208) = 5.1 \times 0.064 \times 0.79 = 0.26\ \mathrm{rad}.$$

(b) According to Table 2.1, the radiation length of Copper is $X_0^{Cu} = 12.9\ \mathrm{g\,cm^{-2}}$. The Copper density is $\rho_{Cu} = 8.96\ \mathrm{g\,cm^{-3}}$. Thus, 2 mm of Copper corresponds to $x = 0.2 \times 8.96 = 1.79\ \mathrm{g\,cm^{-2}}$. The energy of the protons with kinetic energy $T = 200$ MeV is $E = T + m_pc^2 = 1138$ MeV. The corresponding momentum is $pc = \sqrt{E^2 - m_p^2c^4} = 644$ MeV. The relativistic factor $\beta = pc/E = 644/1138 = 0.56$ and $\beta pc = 644 \times 0.56 = 365$ MeV.

The width of the projected angle distribution is:

$$\theta_0^b = \frac{13.6}{365}\sqrt{\frac{1.79}{12.9}}\left(1 + 0.038 \ln\frac{1.79}{12.9}\right) = 0.037 \times 0.37 \times 0.925 = 0.013\ \mathrm{rad}.$$

Problem 2.10 Equation (2.1) depends on three factors: (i) a kinematic factor $(\beta pc)^{-1}$ which does not depend from the characteristic of the target; (ii) the factor z which depends on the particle electric charge; and (iii) the factor $\sqrt{x/X_0}(1 + 0.038 \ln x/X_0)$. This is the only term which depends on the material. The condition of a *thin* target corresponds to $x \ll X_0$. Remember that (Table 2.1) $X_0 = 36$; 24.0; 12.9; 13.8 $\mathrm{g\,cm^{-2}}$ in air, aluminum, copper and iron, respectively. Considering their respective densities, the values correspond to a path length of 300 m in air, 8.9 cm in Al, 1.43 cm in Cu, 1.76 cm in Fe.

Problem 2.11 Let us consider the collision in the system in which the nucleus is at rest. It can be considered that the nucleus mass is $M \simeq m_n A$. The incoming neutron has a mass m_n and its velocity is v, v', before and after the collision, respectively. After the collision, the nucleus has a velocity V. By imposing the non-relativistic energy and momentum conservation laws, one has:

$$\frac{1}{2}m_n v^2 = \frac{1}{2}m_n v'^2 + \frac{1}{2}(m_n A)V^2 \tag{2.7}$$

$$m_n v = m_n v' + (m_n A)V \tag{2.8}$$

The non-relativistic formulae are valid because the neutron kinetic energy T is of the order of a few MeV and $\beta = p/M \simeq \sqrt{\frac{2T}{m_n}} \simeq O(0.1)$.

Solving the system by eliminating V, one obtains a second order equation which admits two solutions. One of the two ($v' = v$) should be eliminated, because it predicts a behavior independent of A corresponding to a non-physical solution. The other solution is:

$$v' = v\frac{1 - A}{1 + A} \tag{2.9}$$

For the different nuclei considered here, one has:

(a) Hydrogen ($A = 1$), $v' = 0$. All the energy of the neutron is transferred to the proton (the small difference in mass between n and p is neglected).
(b) Carbon ($A = 12$), $v' = -0.85v$. The percentage variation in speed (in absolute value) is: $\Delta v/v = |v - v'|/v = 15\%$.
(c) Iron ($A = 56$), $v' = -0.965v$. The percentage variation in speed is: $\Delta v/v = 3.5\%$.

The best moderators are therefore the elements with an atomic number A as small as possible.

References

[2G01] Groom, D.E., Mokhov, N.V., Striganov, S.I.: Muon stopping-power and range tables: 10 MeV–100 TeV. At. Data Nucl. Data Tables **78**, 183–356 (2001). More extensive printable and machine-readable tables are given at http://pdg.lbl.gov/AtomicNuclearProperties/

[2w3] The Internet Seminar (Microscopic World-3 The World of the Atomic Nucleus). http://www.kutl.kyushu-u.ac.jp/seminar/MicroWorld3_E/

Chapter 3
Particle Accelerators and Particle Detection

Problems

3.1. **Energy and momentum**. An on-shell particle with mass m, total energy E and momentum p satisfies the relation $E = +\sqrt{p^2c^2 + m^2c^4}$. For pions with $m = m_{\pi^+} = 140\ \text{MeV}/c^2$, calculate E for $p = 0.1$, 1,10 and 100 GeV/c.
[See solutions]

3.2. **Protons in a magnetic field**. Calculate the curvature radius of the orbit of protons with momentum $p = 10$, 10^3, 10^5 MeV/c in a magnetic field $B =$ 1 Tesla.
[See solutions]

3.3. **Relativistic time dilatation**. Determine the lifetime of a μ^+ with a momentum of 10 GeV/c in the laboratory system; in the muon rest frame $\tau_0 = 2.2\ \mu$s. How far can it go before decaying?
[A: $\gamma = E/m_\mu = 95$; $t = \gamma\tau_0 = 2.1 \cdot 10^{-4}$ s; $d = c\gamma\tau_0 = 6.3 \cdot 10^6$ cm]

3.4. **Center-of-mass energy**. Calculate the energy available in the center-of-mass (c.m.) system using incident pions with 10 GeV/c^2 kinetic energy in the laboratory against
(a) a proton at rest;
(b) an electron at rest.

3.5. **Threshold energies**. Calculate the threshold energy in the laboratory systems for the production of:
(a) π^0 mesons in the reaction $pp \to pp\pi^0$;
(b) π mesons in the reaction $\pi p \to \pi\pi p$;
(c) K^+ mesons in the reaction $pp \to p\Lambda^0K^+$;
(d) Σ^+ hyperons in the reaction $pp \to p\Sigma^+$.

S. Braibant et al., *Particles and Fundamental Interactions: Supplements, Problems and Solutions*, Undergraduate Lecture Notes in Physics,
DOI 10.1007/978-94-007-4135-5_3,

In all cases, assume that the target protons are at rest.

(N.B. The reaction $pp \to p\Sigma^+$ is only possible through the weak interaction.)

[See solutions]

3.6. **Antiproton production-1**. Consider the antiproton production

$$p + p \to pp\overline{p}p$$

through a beam of protons with a momentum $p = 5.5$ GeV/c colliding
(a) on protons in a hydrogen target;
(b) on protons in an iron target.
Calculate the threshold energy and the energy in the c.m. system. Is the production possible in both cases?
[See solutions]

3.7. **Antiproton production-2**.
(a) What is the threshold energy of the reaction $pp \to ppp\overline{p}$?
(b) Determine the kinetic energy available for a target proton bound in a nucleus and moving with a Fermi momentum $\mathbf{p}_F$ (i) towards the incident proton, (ii) in the opposite direction and (iii) orthogonally.
[Hint: see the solution of the previous problem.]

3.8. **Two-body decay**. Consider the decay $\Delta^+ \to \pi^0 p$. Determine the energy and momentum of the two particles in the c.m. system.
[See solutions]

3.9. **Three-body decay**. Consider the decay $K^0 \to \pi^0\pi^+\pi^-$. Determine:
(a) the minimum and maximum values of the π^0 energy and momentum in the K^0 rest system;
(b) the maximum value of the momentum in the lab. system, assuming a K^0 with a momentum $p_K = 100$ GeV/c.
[See solutions]

3.10. ***J/ψ* production**. The J/ψ particle is a meson with a mass $m_{J/\psi} = 3.096$ GeV and made of a $c\overline{c}$ quark-antiquark pair. This particle was discovered (Chap. 9) both in proton-proton collisions and in electron-positron collisions.
(a) A proton beam collides on a target of protons at rest; calculate the incident proton beam energy required for the reaction:

$$pp \to ppJ/\psi$$

(b) In the case of electrons, the particle was discovered in a particle collider in which the e^+ and e^- beams had the same energy but opposite momenta.

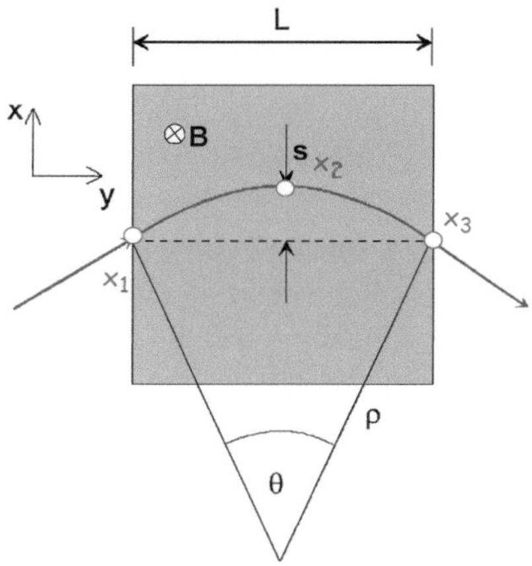

Fig. 3.1 Electrically charged particle moving in a magnetic field **B**. L is the detector dimension, ρ the radius of curvature and s the sagitta of the particle orbit

Calculate the beam energy necessary for the J/ψ production:

$$e^+e^- \to J/\psi$$

The J/ψ decays with a very short lifetime, $\tau \sim 10^{-20}$ s.
[See solutions]

3.11. ***e⁺e⁻* pair**. A positron and an electron produced in a process of pair production in the laboratory system have four-momenta $P_+ = (E_+, \mathbf{p}_+)$ and $P_- = (E_-, \mathbf{p}_-)$. What is the energy of each particle in the system in which the e^+e^- pair has momentum equal to zero?
[See solutions]

3.12. ***pe⁻* collisions**. At the HERA collider in Hamburg, 820 GeV protons collide frontally with 30 GeV electrons.
(a) Calculate the relativistic invariant $\sqrt{s}$.
(b) Guess the reason why the protons have an energy of 820 GeV and the electrons only of 30 GeV.
(c) What would be the c.m. energy using 850 (= 820 + 30) GeV protons colliding with electrons at rest? Why would this situation be less desirable?
[See solutions]

3.13. **Momentum measurement in a magnetic field**. The variables needed for the measurement of a fast electrically charged particle moving in a magnetic field are shown in Fig. 3.1. L is the detector dimension, **B** the (constant) magnetic field assumed perpendicular to the beam direction, ρ the radius of curvature and s the sagitta (see also Fig. 3.9c) of the particle orbit; p_t is the momentum projection in a plane orthogonal to the magnetic field. Neglecting Coulomb multiple scattering, determine:
(a) the relation between p_t and s;

(b) the precision on the measurement of the transverse momentum if $p_t = 1$ GeV/c, $B = 10$ kGauss, $L = 1$ m and $\Delta s = 200$ μm (where Δs is the error on the sagitta).
[See solutions]

3.14. **π, K discrimination**. Consider a magnetic spectrometer which selects positively charged particles with momentum $p = 0.5$ GeV/c in a beam mainly made of π^+ and K^+. The discrimination between the two particles is made using the time-of-flight technique. For this purpose, two scintillation detectors are placed 3 m apart. Each plastic scintillator has a thickness of $\Delta x = 2$ cm, density $\rho = 1.03$ g cm^{-3} and radiation length $X_0 = 40$ cm. Determine:
(a) the π and K velocities;
(b) the energy loss in the first scintillator;
(c) the average deflection angle due to Coulomb multiple scattering for the π^+ and K^+ after the first detector.
[See solutions]

3.15. **Luminosity at the Tevatron**. Calculate the luminosity at the Fermilab Tevatron collider. (a) In the interaction region called B_0, assuming the following parameters: circumference $C = 2\pi R = 6.28$ km; number of protons per bunch $N_p = 6 \cdot 10^{10}$; number of antiprotons per bunch $N_{\overline{p}} = 2 \cdot 10^{10}$; number of bunches $N_B = 6$; correction factor $G = 0.9$, to take into account the finite bunch length of about 50 cm; average transverse radius of each beam $r_{B_0} = 43$ μm. (b) Calculate the luminosity in the interaction region E_0, where the average transverse radius is $r_{E_0} = 380$ μm.
[See solutions]

3.16. **Beam attenuation**. The interactions between particles in the beam pipe of an accelerator with the residual gas present in the beam pipe lead to the attenuation of the beam. This effect reduces the permanence lifetime of particles inside the beam pipe, which must be re-filled.
(a) Calculate the half-life of a proton beam circulating in a storage ring of 100 m radius. The vacuum in the beam pipe corresponds to 10^{-6} mm Hg. Assume that the residual gas is hydrogen and that the total pp cross-section is 40 mb.
(b) Evaluate the beam half-life if the vacuum is 10^{-9} mm Hg.
[See solutions]

3.17. **Reconstruction efficiency**. Cosmic ray tracks can be measured using spark chambers: each chamber has an efficiency of 93%. To reconstruct a track, at least 3 points (and therefore the use of at least 3 chambers) are required. What is the track reconstruction efficiency for a system made of 4 chambers? And for a system of 5 chambers?

3.18. **Synchrotron radiation at LEP**. The LEP collider was a storage ring of 27 km circumference where e^+, e^- beams circulated in opposite direction.
(a) Determine the energy loss due to the synchrotron radiation for each revolution of an electron of 50 GeV energy (LEP-Phase 1) and of 100 GeV energy (LEP-Phase 2).
(b) Determine the intensity of the magnetic field needed to keep on orbit the electrons and positrons for both energies of 50 and 100 GeV; assume a uniform magnetic field along the ring.
(c) Determine the energy loss variation if the e^+, e^- beams are replaced with proton and antiproton beams (assuming they have the same energy as the e^+, e^- beams).
[See solutions]

3.19. **Synchrotron radiation at LHC**. The LEP has been substituted, in the same tunnel, by a pair of storage rings where two proton beams can circulate in opposite directions (the Large Hadron Collider, LHC). The protons are deflected by a set of 1230 magnets, each 14.4 m long. The LHC design foresees that each proton beam shall reach a maximum energy of 7 TeV.
(a) What should be the absolute value of the magnetic field to bend the protons with the maximum energy?
(b) What will be the energy irradiated from a proton per lap; and per second?
[See solutions]

3.20. **Muon factory**. To reduce the energy loss due to the synchrotron radiation of electrons, it was suggested to build a muon storage ring (muon factory). The muons produced in pion decays are collected and directed into the ring. To obtain a high efficiency, it is necessary that the muon direction does not change appreciably from the direction of the charged particle which initially decays.
(a) Determine the muon maximum angle of deflection in the laboratory system with respect to the direction of a pion decaying with a momentum of 20 and 200 GeV/c.
(b) Write the formula of the maximum angle as a function of energy.
[See solutions]

3.21. **Particle-antiparticle pair production**. Consider the production of a particle-antiparticle pair from a high energy positron beam colliding on electrons assumed at rest.
(a) Calculate the minimum positron beam energy necessary to produce a pion pair ($m_\pi = 140$ MeV/c^2).
(b) In the laboratory system, determine the maximum diffusion angle of the pions with respect to the direction of a 150 GeV energy positron beam.
(c) Demonstrate that, for beam energies much larger than the production threshold, the maximum opening angle of the created pair, in the labo-

ratory system, becomes independent of the energy. Calculate the value of this angle for a pair of pions. (If θ_1 is the diffusion angle of the particle and θ_2 that of the antiparticle, the opening angle θ is defined as $\theta = \theta_1 + \theta_2$.)

Supplement 3.1: Synchrotron Radiation

Synchrotron radiation is the electromagnetic radiation generated by charged particles that are accelerated on a curved path or orbit [3W03]. The radiation was named after its discovery in 1946 in a synchrotron accelerator where charged particles were accelerated. The radiated energy is proportional to the fourth power of the particle energy and is inversely proportional to the square of the accelerator radius R. For electron colliders, like LEP, this was the limiting factor on the final beam energy that can be reached.

Classically, any accelerated charged particle emits electromagnetic radiation. The radiated power is given by the relativistic Larmor formula [J99] (in c.g.s. units):

$$P = \frac{dE}{dt} = \frac{2e^2}{3c^3}a^2 \tag{3.1}$$

where a is the particle acceleration. For a non-relativistic circular orbit, the acceleration is just the centripetal acceleration, v^2/R. The relativistic acceleration is obtained from its definition, where $\tau = t/\gamma$ is the proper time and m the particle mass at rest:

$$a = \frac{1}{m}\frac{dp}{d\tau} = \frac{1}{m}\gamma\frac{d(\gamma m v)}{dt} = \gamma^2\frac{dv}{dt} = \gamma^2\frac{v^2}{R} \tag{3.2}$$

The radiated power (neglecting the time dependence of γ) is:

$$\frac{dE}{dt} = \frac{2e^2}{3c^3}a^2 = \frac{2e^2}{3c^3}\left[\gamma^2\frac{v^2}{R}\right]^2 = \frac{2e^2\gamma^4 v^4}{3c^3 R^2} \tag{3.3}$$

Since the velocity $v \to c$ for relativistic particles, the term $\gamma^4 = [E/(mc^2)]^4$ depends on the fourth power of the particle energy and becomes the dominant factor in the energy loss rate. For an accelerator, the radius R is fixed and the synchrotron radiation loss dependence suggests the construction of accelerators as large as possible.

For colliders, the energy loss per orbit can be estimated as:

$$\Delta E = \frac{dE}{dt}\Delta t = \frac{dE}{dt}\left(\frac{2\pi R}{v}\right) = \frac{4\pi e^2\gamma^4\beta^3}{3R} \tag{3.4}$$

Inserting the numerical values for electrons, one has:

$$\Delta E\ [\text{MeV}] = 0.088 \cdot (E\ [\text{GeV}])^4/R\ [\text{m}] \tag{3.5}$$

The LEP e^+e^- collider had a radius of 4300 meters and (during Phase 1) a beam energy of 50 GeV. Thus, for LEP, the energy loss per lap was about $\Delta E_{LEP} =$

$0.088(50)^4/4300 = 128$ MeV, and $\Delta E_{LEP}/E \sim 0.26\%$. It seems a tiny quantity, but the beam frequency was $f = 11240$ Hz (see Problems 3.18 and 3.19).

The energy radiated within the photon energy interval $d(\hbar\omega)$ has a characteristic behavior [C03]. In particular, the critical frequency ω_c is related to the mean energy per photon [P10]:

$$\langle \hbar\omega \rangle = \frac{8}{15\sqrt{3}}\hbar\omega_c \quad \text{where } \omega_c = \frac{3\gamma^3 c}{2R} \tag{3.6}$$

The critical frequency for electrons (and positrons) in a collider is:

$$\hbar\omega_c \text{ [keV]} = 2.22 \cdot (E \text{ [GeV]})^3/R \text{ [m]} \tag{3.7}$$

Synchrotron radiation may also be deliberately produced as radiation source for numerous laboratory applications in the X-ray band. The radiation pattern is distorted by relativistic effects from the isotropic dipole pattern into an extremely forward-pointing radiation cone. This makes circular accelerator the brightest known sources of X-rays. The synchrotron radiation has been used for spectroscopy and diffraction studies since the 1960s. Nowadays, most accelerators dismissed from high energy frontier research are used as storage rings to accelerate electrons which produce synchrotron radiation along different beamlines tangent to the accelerator ring. Typically, the beamline includes X-ray optical devices which focus and collimate the X-rays, and control the photon flux and bandwidth.

Synchrotron light is an ideal tool for many types of research; it also has industrial applications. Some of the advantages of synchrotron radiation are that the X-rays have short wavelength and can penetrate matter and interact with atoms. In addition, these facilities can produce high concentrated, adjustable and polarized radiation thus ensuring a high focusing accuracy even for the smallest targets. Some of the experimental techniques exploiting synchrotron beamlines include: structural analysis of crystalline and amorphous materials; diffraction analysis of microcrystalline samples; crystallography of proteins and other macromolecules; small angle X-ray scattering and absorption spectroscopy; tomography; X-ray imaging in phase contrast mode; high pressure studies, and others.

Solutions

Problem 3.1 We get $E = 0.17$, 1.01, 10.001, 100.0001 GeV, respectively.

Note that at high energies, there is no substantial difference between *momentum* and *energy*; as a consequence, it is therefore convenient to use the natural unit system in which $c = 1$. In this way, mass, energy and momentum are expressed in the same units.

Problem 3.2 One must apply Eq. (3.20b). For 10 MeV/c protons, one has:

$$R(m) = p \text{ (GeV/c)}/(0.3 \cdot B \text{ (Tesla)}) = 3.3 \times 10^{-2} \text{ m.}$$

Since the formula depends linearly on the momentum, the curvature radius is respectively 10^2, 10^4 times larger for the two other cases.

Problem 3.5 Consider the relativistic invariant, $E^2 - p^2$ ($c = 1$), both in the c.m. system and in the laboratory system:

$$E^{*2} - p^{*2} = E_{lab}^2 - p_{lab}^2. \tag{3.8}$$

In the c.m. system, one has: $p^{*2} = 0$. In the laboratory system, the incident proton (1) collides on the target proton (2) at rest. One has: $E_{lab} = (E_1 + m_2)$, $p_{lab} = p_1$; therefore, expressing the particle energy as the sum of the kinetic energy and the mass energy, $E_1 = T_1 + m_1$, one can write:

$$\begin{aligned} E^{*2} &= (E_1 + m_2)^2 - p_1^2 = E_1^2 + m_2^2 + 2E_1 m_2 - p_1^2 \\ &= m_1^2 + m_2^2 + 2(T_1 + m_1)m_2 = (m_1 + m_2)^2 + 2T_1 m_2 \end{aligned} \tag{3.9}$$

From this and using $s = E^{*2}$, one has:

$$T_1 = \frac{s - (m_1 + m_2)^2}{2m_2} \underset{p+p}{\Longrightarrow} \frac{s - 4m_p^2}{2m_p} \tag{3.10}$$

(in the last equality, the incident and target particles are assumed to be protons with $m_1 = m_2 = m_p$).

(a) For $pp \to pp\pi^0$, the minimum energy required to produce the 3 final state particles is equal to $\sqrt{s} = 2m_p + m_{\pi^0}$. In the c.m. system, each proton must have, in addition to their rest mass, a kinetic energy equal to at least half the mass of the third particle (the pion in this case) that must be produced (here, about 67.5 MeV).

In the lab. system, from Eq. (3.10):

$$\begin{aligned} T_1 &= \frac{(2m_p + m_{\pi^0})^2 - 4m_p^2}{2m_p} = \frac{4m_p^2 + m_{\pi^0}^2 + 4m_p m_{\pi^0} - 4m_p^2}{2m_p} \\ &= \frac{m_{\pi^0}^2 + 4m_p m_{\pi^0}}{2m_p} \underset{(m_\pi \ll m_p)}{\Longrightarrow} 2m_{\pi^0} \simeq 270 \text{ MeV} \end{aligned}$$

(b) For the $\pi p \to \pi\pi p$ reaction, one has:

$$T_1 = \frac{(2m_\pi + m_p)^2 - (m_\pi + m_p)^2}{2m_p} = m_\pi \left[\frac{3}{2}\frac{m_\pi}{m_p} + 1\right] = 1.22 m_\pi = 171 \text{ MeV}$$

(c) For the $pp \to p\Lambda^0 K^+$ reaction, one has:

$$\begin{aligned} T_1 &= \frac{(m_p + m_\Lambda + m_K)^2 - 4m_p^2}{2m_p} = \frac{(938 + 1115 + 497)^2 - 4 \cdot 938^2}{2 \cdot 938} \\ &= 1590 \text{ MeV} \end{aligned}$$

(d) For the $pp \to p\Sigma^+$ reaction, one has:

$$T_1 = \frac{(m_p + m_\Sigma)^2 - 4m_p^2}{2m_p} = \frac{(938 + 1190)^2 - 4 \cdot 938^2}{2 \cdot 938} = 538 \text{ MeV}$$

Problem 3.6

(a) Production on hydrogen (protons): This is an "economic" reaction in terms of the baryon number conservation: $1+1 \to 1+1+(-1)+1$.
The threshold energy corresponds to $E_t = 4 \cdot m_p = 4 \cdot 0.938 = 3.75$ GeV.
The incident proton energy is $E_1 = \sqrt{m_p^2 + p_p^2} = \sqrt{0.938^2 + 5.5^2} = 5.58$ GeV.
The square of the sum of the four-momenta (P_1 for the incoming proton, and P_2 for the proton at rest) is a relativistic invariant. The total energy in the c.m. system corresponding to the given values is:

$$\begin{aligned} s &= |P_1 + P_2|^2 = \left|(E_1, \mathbf{p}_1) + (m_p, 0)\right|^2 = \left|(E_1 + m_p, \mathbf{p}_1)\right|^2 \\ &= E_1^2 + m_p^2 + 2E_1 m_p - \mathbf{p}_1^2 = 2m_p^2 + 2E_1 m_p = 2m_p(m_p + E_1) \\ \sqrt{s} &= \sqrt{2m_p(m_p + E_1)} = 3.48 \text{ GeV} \end{aligned}$$

This value $\sqrt{s} = 3.48$ GeV is smaller than the threshold energy $E_t = 3.75$ GeV. Therefore, the energy available in the c.m. system is insufficient to create the desired final state.

(b) Production on iron nuclei: the target protons have a Fermi momentum of about $p_F \simeq 200$ MeV/c in absolute value (see Chap. 14). The energy of the target proton is, in the most favorable case, $E_2 = \sqrt{0.938^2 + 0.2^2} = 0.959$ GeV. The total energy in the c.m. system is equal to:

$$\begin{aligned} s &= (E_1 + E_2)^2 - (\mathbf{p}_1^2 + \mathbf{p}_2^2)^2 \\ &= 2m_p^2 + 2E_1 E_2 - 2\mathbf{p}_1 \mathbf{p}_2 \\ &= 2\left(m_p^2 + E_1 E_2 + p_1 p_2\right) \\ \sqrt{s} &= \sqrt{2\left(m_p^2 + E_1 E_2 + p_1 p_2\right)} \\ \sqrt{s} &= \sqrt{2\left(m_p^2 + 5.58 \cdot 0.959 + 5.5 \cdot 0.2\right)} = 3.83 \text{ GeV} \end{aligned}$$

$\sqrt{s} = 3.83$ GeV is above the threshold energy $E_t = 3.75$ GeV. Therefore, the energy available in the c.m. system can be sufficient to create the final state with an antiproton.

Problem 3.8 To maintain generality, the decaying particle is denoted A and the final state particles are denoted B, C. Their four-momenta are respectively $(m_A, 0)$ and $(E_B^*, \mathbf{p}_B^*)$, $(E_C^*, \mathbf{p}_C^*)$. The index * reminds us that the variables are defined in the c.m. system (for a two-body decay, it coincides with the frame where the particle A is at rest). Before the decay, one has $\sqrt{s} = m_\Delta = 1232$ MeV.

After the decay, one has $(E_B^* + E_C^*, \mathbf{p}_B^* + \mathbf{p}_C^*)$ and:

$$\begin{aligned} s &= m_B^2 + m_C^2 + 2E_B^* \cdot E_C^* - 2\mathbf{p}_B^* \mathbf{p}_C^* \\ &= m_B^2 + m_C^2 + 2E_B^* \cdot \left(\sqrt{s} - E_B^*\right) + 2p_B^{*2} \\ &= m_B^2 + m_C^2 - 2m_B^2 + 2E_B^* \cdot \sqrt{s} \end{aligned}$$

having used the fact that $\sqrt{s} = E_B^* + E_C^*$, $E_B^{*2} - p_B^{*2} = m_B^{*2}$ and that $\mathbf{p}_B^* = -\mathbf{p}_C^*$. Finally, one can write:

$$E_B^* = \frac{s - m_C^2 + m_B^2}{2\sqrt{s}} \tag{3.11}$$

and similarly

$$E_C^* = \frac{s - m_B^2 + m_C^2}{2\sqrt{s}} \tag{3.12}$$

Placing the numerical values in the equations, one finds: $E_B^* = E_\pi^* = \frac{1232^2 - 938^2 + 140^2}{2 \cdot 1232} = 267$ MeV and $E_C^* = E_p^* = \frac{1232^2 - 140^2 + 938^2}{2 \cdot 1232} = 965$ MeV. The corresponding values of the absolute values of the momenta are: $p_p^* = p_\pi^* = \sqrt{E_\pi^{*2} - m_\pi^2} = 227$ MeV/c.

Problem 3.9 This problem differs from the previous one since there are now three particles in the final state (B, C, D), which can share in many ways the rest mass energy of the decaying particle: $\sqrt{s} = m_A = E_B^* + E_C^* + E_D^*$. In this problem, one has $m_A = m_{K^0} = 498$ MeV.

(a) We can now identify $B = \pi^0$. The minimum energy corresponds to the production of a π^0 at rest, with $E_{\pi^0}^{*,min} = m_\pi^0 = 135$ MeV.
The maximum energy is obtained when the momenta of the two charged pions are parallel but with a direction opposite to that of the π^0. Using Eq. (3.11) and the same calculation as that of the previous problem, one has ($m_{\pi^+} = m_{\pi^-} = 140$ MeV, $m_{\pi^0} = 135$ MeV):

$$E_{\pi^0}^{*,max} = \frac{s^2 - (m_{\pi^+} + m_{\pi^-})^2 + m_{\pi^0}^2}{2\sqrt{s}} = 189 \text{ MeV} \tag{3.13}$$

The corresponding maximum momentum is: $p_{\pi^0}^{*,max} = \sqrt{189^2 - 135^2} = 132$ MeV/c.

(b) The Lorentz γ factor for the K^0 (the *Lorentz boost*) must be first calculated ($E_K = p_K$):

$$\gamma = \frac{E_K}{m_{K^0}} = 100/0.498 = 201; \qquad \beta \simeq 1$$

In the laboratory system, the maximum momentum is obtained when the π^0 is emitted in the direction opposite to the $\pi^+\pi^-$ system but in the same direction as the K, i.e., with $\cos\theta^* = +1$. Using the Lorentz transformation:

$$p_{\pi^0}^{max} = \gamma \left(p_{\pi^0}^{*,max} \cos\theta^* + \beta E_{\pi^0}^{*,max} \right) = 201(0.132 + 0.189) = 64.5 \text{ GeV/c}.$$

Problem 3.10 The situations presented in this problem correspond to the typical production of a particle in the *laboratory system* (case (a)) or in the *center-of-mass system* (case (b)).

(a) The energy-momentum four-vectors of the two protons are: $(E, \mathbf{p}_1)$, $(m_p, 0)$. The variable E is the incident proton total energy that must be determined. Calculating s in the laboratory system, one has:

$$s = (E + m_p, \mathbf{p}_1)^2 = \left(E^2 - p_1^2\right) + m_p^2 + 2Em_p = 2m_p^2 + 2Em_p \tag{3.14}$$

Solving the equation in terms of E, one has:

$$E = \frac{s - 2m_p^2}{2m_p} \tag{3.15}$$

We must now determine the quantity s of the final state (two protons plus the J/ψ particle). The minimum value of the invariant s corresponds to the production at rest:

$$\begin{aligned} s &= \left[(m_{J/\psi}, 0) + (m_p, 0) + (m_p, 0)\right]^2 = (m_{J/\psi} + m_p + m_p)^2 \\ &= 4m_p^2 + m_{J/\psi}^2 + 4m_p m_{J/\psi} \end{aligned} \tag{3.16}$$

Placing Eq. (3.16) in Eq. (3.15) and using the numerical values $m_p = 0.938$ GeV, $m_{J/\psi} = 3.096$ GeV, one obtains:

$$E = \frac{2m_p^2 + m_{J/\psi}^2 + 4m_p m_{J/\psi}}{2m_p} = 12.24 \text{ GeV} \tag{3.17}$$

(b) In the case of a production in a collider, the calculation of the invariant quantity s is directly obtained in the c.m. system since the lepton-antilepton pair has four-momenta: $(E, \mathbf{p})$, $(E, -\mathbf{p})$. In this case, one has:

$$s = (2E)^2 = m_{J/\psi}^2 \tag{3.18}$$

from which one can derive that the electron and positron energy in the collider must be:

$$E = \frac{m_{J/\psi}}{2} = 1.548 \text{ GeV} \tag{3.19}$$

Problem 3.11

$$\begin{aligned} s &= (E_+ + E_-, \mathbf{p}_+ + \mathbf{p}_-)^2 = E_+^2 + E_-^2 + 2E_+E_- - \left(p_+^2 + p_-^2 + 2p_+p_- \cos\theta\right) \\ &= 2m_e^2 + 2E_+E_- - 2p_+p_- \cos\theta \end{aligned}$$

In the system frame where the four-momenta are $(E'_+, \mathbf{p}')$, $(E'_-, -\mathbf{p}')$, one has $E'_+ = E'_- = E'$ and $s = (2E')^2$ (see Eq. (3.18)). Therefore:

$$E' = \sqrt{\frac{2m_e^2 + 2E_+E_- - 2p_+p_- \cos\theta}{2}}$$

Problem 3.12

(a) The relativistic energy-momentum invariant is equal to:

$$s = m_1^2 + m_2^2 + 2E_1E_2 - 2p_1p_2\cos\theta \tag{3.20}$$

In a high-energy collider, $\cos\theta = -1$ and: $|\mathbf{p}_1| \simeq E_1 \gg m_1$, $|\mathbf{p}_2| \simeq E_2 \gg m_2$. Equation (3.20) becomes:

$$s = m_1^2 + m_2^2 + 2E_1E_2 + 2p_1p_2 \simeq 4E_1E_2 \tag{3.21}$$

$$\sqrt{s} \simeq 2\sqrt{E_1E_2} = 2\sqrt{820 \cdot 30} = 314 \text{ GeV}. \tag{3.22}$$

(b), (c) Particles lose energy through synchrotron radiation with an energy dependence $(E/m)^4$, see Supplement 3.1; to minimize the energy loss the best option is to use relatively low-energy electrons.
Indeed, for electrons at rest and protons of 850 GeV, one has:

$$s = m_1^2 + m_2^2 + 2m_2E_1$$
$$\sqrt{s} \simeq \sqrt{2m_2E_p + m_1^2} = \sqrt{2m_eE_p + m_p^2} \tag{3.23}$$
$$\simeq \sqrt{2 \cdot 850 \cdot 0.511 \cdot 10^{-3} + 0.938^2} = 1.32 \text{ GeV}.$$

In this case, the energy available in the c.m. is much smaller than (3.22).

Problem 3.13

(a) The track curvature and the sagitta are assumed to be measured with a tracking device (for instance, the central device in collider detectors, see Chap. 9). The momentum component in the plane transverse to the magnetic field is (see Problem 3.2):

$$p_t \text{ [GeV]} = 0.3B \text{ [T]}\rho \text{ [m]}$$

Referring to the geometry sketched in Fig. 3.1, one can write:

$$\frac{(L/2)}{\rho} = \sin\frac{\theta}{2} \simeq \frac{\theta}{2} \text{ (for small } \theta) \rightarrow \theta \simeq \frac{L}{\rho} = \frac{0.3BL}{p_t} \tag{3.24}$$

and:

$$s = \rho\left(1 - \cos\frac{\theta}{2}\right) \simeq \rho\left[1 - \left(1 - \frac{1}{2}\frac{\theta^2}{4}\right)\right] = \rho\frac{\theta^2}{8} \simeq \frac{0.3}{8}\frac{L^2B}{p_t} \tag{3.25}$$

(b) Assuming that $p_t = 1$ GeV/c, $L = 1$ m and $B = 10$ kGauss $= 1$ T, the sagitta is:

$$s = \frac{0.3}{8}\frac{1^2 1}{1} = 0.038 \text{ m} = 3.8 \text{ cm}$$

Neglecting the uncertainties in L and B, the relative uncertainty on p_t is:

$$\frac{\Delta p_t}{p_t} = \frac{\Delta s}{s} = \frac{200\ \mu\text{m}}{0.038} = 5 \times 10^{-3} = 0.5\%$$

Problem 3.14

(a) Let us use the natural units ($c = 1$). The particle velocity can be determined from the particle momentum and energy:

$$p = mv\gamma; \qquad E = m\gamma \rightarrow v = \frac{p}{E}$$

The masses of the two particles are $m_\pi = 140$ MeV and $m_K = 494$ MeV. Thus, for $p = 0.5$ GeV (inserting the factor c at the end):

$$v_\pi = \frac{p}{\sqrt{p^2 + m_\pi^2}} = 0.963\ c$$

$$v_K = \frac{p}{\sqrt{p^2 + m_K^2}} = 0.711\ c$$

The time-of-flights between the two scintillator layers are:

$$\Delta t_\pi = \frac{3\ \text{m}}{v_\pi} = \frac{10}{0.963} \times 10^{-9}\ \text{s} = 10.4\ \text{ns}$$

$$\Delta t_K = \frac{3\ \text{m}}{v_K} = \frac{10}{0.711} \times 10^{-9}\ \text{s} = 14.1\ \text{ns}$$

(b) The energy loss in the first scintillator layer depends on the particle $\beta\gamma$. The value $\beta\gamma = p/m$ for the two particles is:

$$(\beta\gamma)_\pi = (0.5/0.14) = 3.7; \qquad (\beta\gamma)_K = (0.5/0.5) = 1$$

By inspection of the curve for the energy loss in carbon shown in Fig. 2.2a, we get $dE/dx \sim 1.75$ MeV g^{-1} cm^2 in the case of the pion, and $dE/dx \sim 2.7$ MeV g^{-1} cm^2 for the kaon. We use the carbon curve as representative of a material made of light elements, as a plastic scintillator, which is a polymer with C, O and H nuclei. The energy loss in the detector thickness is:

$$(\Delta E)_\pi = (dE/dx) \times \rho \times \Delta x = 1.75 \times 1.03 \times 2 = 3.6\ \text{MeV}$$

$$(\Delta E)_K = (dE/dx) \times \rho \times \Delta x = 2.7 \times 1.03 \times 2 = 5.6\ \text{MeV}$$

(c) The average deflection angle due to Coulomb multiple scattering is given by Eq. (2.1). For the kaon, with $\beta_K = 0.711$:

$$\theta_0^K = \frac{13.6\ \text{MeV}}{\beta_K cp}\sqrt{\frac{\Delta x}{X_0}}[1 + 0.038 \ln(\Delta x/X_0)]$$

$$= \frac{13.6\ \text{MeV}}{0.71 \times 500\ \text{MeV}}\sqrt{\frac{2}{40}}[1 - 0.11] = 7.5 \times 10^{-3}\ \text{rad}$$

For the pion, $\theta_0^\pi = 5.5 \times 10^{-3}$ rad.

Problem 3.15

(a) The formula for the luminosity in terms of the collider parameters is (Sect. 3.3):

$$\mathcal{L} = \frac{f N_p N_{\overline{p}} N_B G}{4\pi r_{B0}^2}$$

$$\begin{cases} \text{Revolution period} & \tau = 2\pi R/c = 6.28 \cdot 10^3/3 \cdot 10^8 = 20.93\ \mu\text{s} \\ \text{Revolution frequency} & f = 1/\tau = 1/2.093 \cdot 10^{-5} = 4.78 \cdot 10^4\ \text{Hz} \end{cases} \tag{3.26}$$

With the data given in the problem, the luminosity in the region B_0 is:

$$\mathcal{L}_{B0} = \frac{4.78 \cdot 10^4 \times 6 \cdot 10^{10} \times 2 \cdot 10^{10} \times 6 \times 0.9}{4\pi (4.3 \cdot 10^{-3}\ \text{cm})^2} \simeq 1.3 \cdot 10^{30}\ \text{cm}^{-2}\,\text{s}^{-1}.$$

The denominator of Eq. (3.26), $A = 4\pi r_{B0}^2$, contains the beam mean radius, r_{B0}, which is actually larger horizontally than vertically. More accurate computations take account of this distortion.

(b) In E_0, the luminosity in B_0 is reduced by the factor $(r_{E0}/r_{B0})^2 = (380/43)^2 = 78$. Therefore, one has:

$$\mathcal{L}_{E0} = \mathcal{L}_{B0}/78 \simeq 1.7 \times 10^{28}\ \text{cm}^{-2}\,\text{s}^{-1}.$$

Problem 3.16 In technical units, the *vacuum* inside a container is expressed in mm Hg. Remember that one atmosphere corresponds to 760 mm Hg. Actually, the interesting parameter for the propagation of particles is the *density number* of the residual gas nuclei inside the vacuum pipe. This can be estimated from the well known equation for gases:

$$\frac{P_{atm}}{P} = \frac{\rho_{atm}}{\rho} \tag{3.27}$$

where the atmospheric density corresponds to $\rho_{atm} = 1.22\ \text{kg}\,\text{m}^{-3} = 1.22 \cdot 10^{-3}\ \text{g}\,\text{cm}^{-3}$ and $P_{atm} = 760$ mm Hg.

The density of matter inside the vacuum pipe, where the pressure corresponds to $P = 10^{-6}$ mm Hg is therefore:

$$\rho = \rho_{atm}\left(\frac{P}{P_{atm}}\right) = \rho_{atm}\left(\frac{10^{-6}}{760}\right) = 1.3 \times 10^{-9} \rho_{atm} = 1.6 \times 10^{-12}\ \text{g}\,\text{cm}^{-3}$$

The density number N_a (cm^{-3}) of the gas inside the pipe can be derived from ρ, assuming that the residual gas is composed of hydrogen with $A = 1$, through the relation (N_A is the Avogadro's number):

$$N_a = \frac{\rho N_A}{A} = 6.0 \times 10^{23} \times 1.6 \times 10^{-12} = 9.6 \times 10^{-11}\ \text{cm}^{-3} \tag{3.28}$$

The beam is attenuated by following the exponential law $N(x) = N_0 e^{-x/\lambda}$, where λ is the *interaction length*. λ is inversely proportional to the cross-section σ (cm^2) and to the density number (cm^{-3}):

$$\lambda = \frac{1}{\sigma N_a} = \frac{1}{40 \cdot 10^{-27} \times 9 \cdot 10^{-11}} = 2.6 \times 10^{13}\ \text{cm} = 2.6 \times 10^{11}\ \text{m} \tag{3.29}$$

The number of revolutions in the accelerator, N_λ, which corresponds to λ is:

$$N_\lambda = \frac{2.6 \times 10^{11} \text{ m}}{2\pi R} = 4.1 \times 10^8$$

where $R = 100$ m is the accelerator radius. To cover one revolution, a particle, travelling at the light speed c, takes:

$$\Delta t = \frac{2\pi R}{c} = \frac{628}{3 \times 10^8} = 2.1 \times 10^{-6} \text{ s}$$

The beam lifetime τ in the vacuum pipe (i.e., the time when the remaining particles are $1/e$ of the initial ones) is the number of covered revolutions N_λ multiplied by the time necessary to cover one revolution:

$$\tau = N_\lambda \times \Delta t = 4.1\ 10^8 \times 2.1 \cdot 10^{-6} = 860 \text{ s}$$

In the problem, it is also requested to compute the beam half-life $t_{1/2}$, i.e., the time when 50% of the protons survive in the vacuum pipe. This quantity is connected to the lifetime given in Eq. (14.20):

$$t_{1/2} = 0.693\tau \simeq 600 \text{ s} = 10 \text{ min}$$

(b) For $P = 10^{-9}$ mm Hg, the half-life is three orders of magnitude higher, i.e., about 170 hours.

Problem 3.18

(a) The energy loss per orbit is given in Eq. (3.4):

$$\Delta E = \frac{4\pi e^2 \gamma^4 \beta^3}{3R}$$

Inserting the numerical values, one has: ΔE [MeV] $= 0.088 \cdot (E$ [GeV]$)^4/$ R [m]. For $R = 4300$ m, it corresponds to:

$$\Delta E(\text{LEP-phase I}) = 0.088 \cdot 50^4/4300 = 128 \text{ MeV}$$

$$\Delta E(\text{LEP-phase II}) = 0.088 \cdot 100^4/4300 = 2046 \text{ MeV}$$

Increasing the beam energy by a factor of 2 causes an increase of the radiation energy loss by a factor of 16.

(b) Using the result of Problem 3.2, at high energy, one has $E = pc$ and one finds B [T] $= E$ [GeV]$/(0.3R$ [m]$)$ from which $B = 0.04$ T at $E = 50$ GeV and $B = 0.08$ T at $E = 100$ GeV.

(c) The radiation energy loss depends on $\gamma^4 = (E/m)^4$. At the same energy, the emission from a proton beam is a factor $(m_e/m_p)^4 = 9 \times 10^{-14}$ smaller than that of an electron beam with the same energy.

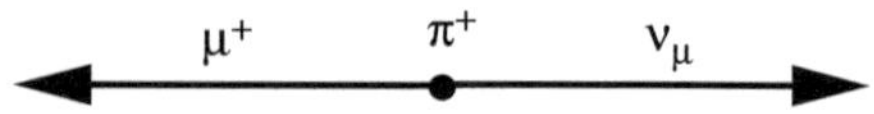

Fig. 3.2 Pion decay in the c.m. system

Problem 3.19

(a) If the magnets were evenly distributed throughout the tunnel of radius $R = 4300$ m, the magnetic field (see previous problem) for $E = 7$ TeV should be B [T] $= E$ [GeV]/(0.3R [m]) $= 5.4$ T. However, the magnets occupy only [14.4 m $\times$ 1230/(2π $\times$ 4300)] $= 66\%$ of the circumference. Each dipole magnetic field should therefore have $B = 5.4/0.66 = 8.3$ T.
(b) The proton revolution period is $\tau = 2\pi R/c = 9 \cdot 10^{-5}$ s. Using the expression for the energy loss for electrons, scaled by $(m_e/m_p)^4$, one obtains:

$$\Delta E(\text{LHC}) = 0.088 \cdot \frac{7000^4}{4300} \cdot \left(\frac{m_e}{m_p}\right)^4 = 4.4 \times 10^{-3}\ \text{MeV} = 4.4\ \text{keV}$$

The energy loss per second is: $\frac{\Delta E(\text{LHC})}{\Delta t} = 4.4 \times 10^{-3}/9 \cdot 10^{-5} = 49$ MeV/s.

Problem 3.20 Let us first consider the pion decay (Fig. 3.2) in the c.m. system ($m_\pi = 139.6$ GeV; $m_\mu = 105.7$ GeV). The muon energy E^*_μ in the c.m. system has been determined in Problem 2.7:

$$E^*_\mu = \frac{s + m^2_\mu}{2\sqrt{s}} = \frac{m^2_\pi + m^2_\mu}{2m_\pi} = 109.8\ \text{MeV}$$

and therefore, the muon momentum is $p^*_\mu = \sqrt{E^{*2}_\mu - m^2_\mu} = 29.8$ MeV/c. Similarly for the neutrino, considering that in the c.m. system $E^*_\mu + E^*_\nu = m_\pi$, one has:

$$E^*_\nu = \frac{s - m^2_\mu}{2\sqrt{s}} = \frac{m^2_\pi - m^2_\mu}{2m_\pi} = 29.8\ \text{MeV}$$

To calculate the emission angle in the laboratory system, it is necessary to determine the longitudinal $p_\parallel$ and transverse $p_\perp$ components of the muon momentum. The emission angle is given by the relation:

$$\tan\theta = \frac{p_\perp}{p_\parallel} \tag{3.30}$$

The longitudinal and transverse components of the muon momentum are determined by the Lorentz transformation:

$$p_\parallel = \gamma\left(\beta E^*_\mu + p^*_\mu \cos\theta^*\right) \tag{3.31}$$

$$p_\perp = p^{*\mu}_\perp = p^*_\mu \sin\theta^* \tag{3.32}$$

where θ^* is the muon emission angle in the rest system of the pion with respect to the direction of the pion in the laboratory system (this corresponds to the direction of the Lorentz boost). The value of the Lorentz boost γ is determined by the pion energy:

$$\gamma = \frac{E_\pi}{m_\pi} = 143 \quad \text{for } E_\pi = 20\ \text{GeV}$$

The $\beta = \sqrt{1-\gamma^{-2}}$ value can therefore be obtained, but in this context, it can be approximated with $\beta = 1$. From Eq. (3.30), and recalling that $p^*_\mu = 29.8$ MeV, one obtains:

$$\begin{aligned}\tan\theta &= \frac{p_\perp}{p_\parallel} = \frac{p^*_\mu \sin\theta^*}{\gamma(\beta E^*_\mu + p^*_\mu \cos\theta^*)} \\ &= \frac{29.8\sin\theta^*}{\gamma(109.8 + 29.8\cos\theta^*)} = \frac{f(\theta^*)}{\gamma}\end{aligned} \tag{3.33}$$

The function $f(\theta^*)$ does not depend on the particular kinematic situation. The maximum value (maximizing the function $f(\theta^*)$ or representing it in a graphical way and finding the maximum by inspection), is $f^{max}(\theta^*) = 0.255$ and it occurs at $\theta^* = 1.85$ rad = 106° (note that this angle is larger than 90°). In the lab. system, this corresponds to

$$\theta_{max} = \arctan\left(\frac{0.255}{\gamma}\right) \simeq \frac{0.255}{\gamma} = 1.8 \cdot 10^{-3}\ \mathrm{rad} = 0.10°$$

for $\gamma = 143$ corresponding to pions of 20 GeV/c momentum. As the Lorentz boost factor is linearly dependent on the energy, θ_{max} is one order of magnitude smaller for pions of 200 GeV/c.

References

[3W03] Wiedemann, H.: Synchrotron Radiation. Springer, Berlin (1994). ISBN: 978-3-540-43392-7

Chapter 4
The Paradigm of Interactions: The Electromagnetic Case

Problems

4.1. **Yukawa range of weak interactions**. The weak interaction is due to the exchange of $W^{\pm}$ (or Z^0) bosons. In the framework of a Yukawa-like model describing the weak interaction, and assuming that the W has a mass of $M = 80$ GeV/c^2, calculate the range of the weak interaction and compare the obtained value with the size of a nucleon.
[A: $R = \hbar/Mc \simeq 3 \times 10^{-16}$ cm]

4.2. **Free electron emission**. Show that the decay $e \to e\gamma$ (Fig. 4.2a) cannot satisfy, at the vertex, the energy and momentum conservation laws if the initial electron and the particles in the final state are free (this process is therefore prohibited).
[See solutions]

4.3. **Impact parameter and elastic scattering angle**. Using classical kinematics, demonstrate, for the Coulomb elastic scattering (see Fig. 4.9b), the relation $\mathrm{tg}\,\frac{\theta}{2} = \frac{zZe^2}{2E_c b}$ between the scattering angle θ, the kinetic energy $E_c = (p^2/2m)$ of the incident particle and the impact parameter b.

4.4. **Lifetime and path length**. A π^- meson interacts in a bubble chamber, in a point of coordinates $x_1 = -50$ cm, $y_1 = 50$ cm, $z_1 = 20$ cm, $t_1 = 0$ s producing some mesons, including a K^- (vertex 1). The K^- meson travels at a constant speed until it interacts in a region containing counters and chambers (vertex 2 at the position $x_2 = 88$ cm, $y_2 = 48$ cm, $z_2 = 25$ cm, $t_2 = 5.31$ ns). Calculate for the K^-:
(a) the travelled distance,
(b) the flight time (corresponding to its lifetime measured in the laboratory system),
(c) the velocity,

S. Braibant et al., *Particles and Fundamental Interactions: Supplements, Problems and Solutions*, Undergraduate Lecture Notes in Physics,
DOI 10.1007/978-94-007-4135-5_4, © Springer Science+Business Media Dordrecht 2012

(d) the K^- lifetime in the frame in which the particle is at rest,
(e) the probability that the K^- decays along the travelled distance.
[See solutions]

4.5. **Radiocarbon dating**. The half-life, $t_{1/2}$, of the carbon 14 isotope, ${}^{14}_{6}\mathrm{C}$, is 5730 years. The concentration of this isotope, in the atmosphere, with respect to the stable isotope ${}^{12}_{6}\mathrm{C}$ is ${}^{14}_{6}\mathrm{C}/{}^{12}_{6}\mathrm{C} = 1.0 \cdot 10^{-12}$.
(a) What is the ratio ${}^{14}_{6}\mathrm{C}/{}^{12}_{6}\mathrm{C}$ in an object which ended its life 12000 years ago?
(b) An artifact made of wood has a concentration of ${}^{14}_{6}\mathrm{C}$ corresponding to 58% of the concentration in a similar object made of freshly cut wood. Determine the age of the original sample.
[See solutions]

4.6. **Rutherford scattering**. The α particles are helium nuclei. A flux of $\Phi = 5.0 \cdot 10^7$ particles/s of 8.0 MeV energy collides on a gold target 4.0 μm thick. A sector of a circular detector, concentric with the beam, is placed at a distance of 3.0 cm after the target (see Fig. 4.9a). The smaller radius of the circular sector from the beam axis is $r_{min} = 5.0$ mm while the larger radius is $r_{max} = 7.0$ mm. Determine the number of particles reaching the detector per second. Discuss and comment the case in which $r_{min} \to 0$.
[See solutions]

4.7. **Electromagnetic transition probability**. Using the second Fermi golden rule, Eq. (4.28), show that the annihilation probability of positronium (bound state e^+e^-) in the singlet state ${}^1S_0$ (i.e., the state in which the electron and positron spins are antiparallel) is given by:

$$W({}^1S_0 \to \gamma\gamma) = \frac{4\pi\alpha^2}{m_e^2}|\psi(0)|^2 \tag{4.1}$$

where $|\psi(0)|^2$ represents the probability that the electron and the positron are in the same region of space. Draw the Feynman diagram of the annihilation process into two photons.
[See solutions]

4.8. **Positronium annihilation**. In the hydrogen atom, the wave function of the state ${}^1S_0$ of the electron moving around the proton, is known from atomic physics:

$$|\psi(r)| = \frac{1}{\sqrt{\pi}a^{3/2}}e^{-r/a} \tag{4.2}$$

where a is the Bohr radius of the hydrogen atom:

$$a = \frac{4\pi\epsilon_0\hbar^2}{m_e e^2} = \frac{\hbar}{m_e c\alpha} = 0.5 \cdot 10^{-10}\ \text{m} \tag{4.3}$$

Using Eq. (4.1), determine the lifetime of the positronium in the $^1S_0$ state.
[See solutions]

4.9. **Dirac theory-I.**[1] Dirac proposed an equation that would satisfy the energy-momentum relativistic relation ($E^2 = p^2 + m^2$), but containing only first order space-time derivative. Show that the eigenvalue equation:

$$E\psi = i\hbar\frac{\partial}{\partial t}\psi = H\psi \tag{4.4}$$

can be written as:

$$i\hbar\gamma^0\frac{\partial}{\partial t}\psi + i\hbar\boldsymbol{\gamma}\cdot\nabla\psi - m\psi = 0 \tag{4.5}$$

where the γ matrices are defined in Appendix A4.
[See solutions]

4.10. **Dirac theory-II**. Prove that the Dirac equation Hamiltoninan:

$$H = \boldsymbol{\alpha}\cdot\mathbf{p}c + \beta mc^2 \tag{4.6}$$

commutes with the total angular momentum operator

$$\mathbf{J} = \hbar\mathbf{L} + \frac{1}{2}\hbar\boldsymbol{\sigma}. \tag{4.7}$$

[See solutions]

4.11. **Dirac theory-III**. Prove that the Dirac equation Hamiltonian:

$$H = \boldsymbol{\alpha}\cdot\mathbf{p}c + \beta mc^2 \tag{4.8}$$

commutes with the generalized helicity operator (A.27):

$$\Lambda = \begin{pmatrix} \frac{\boldsymbol{\sigma}\cdot\mathbf{p}}{p} & 0 \\ 0 & \frac{\boldsymbol{\sigma}\cdot\mathbf{p}}{p} \end{pmatrix} \tag{4.9}$$

[See solutions]

[1] For the following four problems, the reader should know the content of Appendix A4.

4.12. **Dirac theory-IV**. Show that the 4-component wave function

$$\psi = \begin{pmatrix} \phi \\ \chi \end{pmatrix} e^{i(\mathbf{p}\cdot\mathbf{r}-Et)} \tag{4.10}$$

where ϕ and χ are 2-component spinors, satisfies the Dirac equation if:

$$\phi = \frac{\boldsymbol{\sigma}\cdot\mathbf{p}}{E-m}\chi \quad \text{and} \quad \chi = \frac{\boldsymbol{\sigma}\cdot\mathbf{p}}{E+m}\phi. \tag{4.11}$$

[See solutions]

Supplement 4.1: Radiocarbon Dating

The radiocarbon dating is one of the most famous dating techniques which uses radioactive isotopes. Radiocarbon is quite important because it can be used to determine the age of matter which lived up to $\sim 50,000$ y ago [4B90].

Cosmic rays consist of high energy protons and heavier nuclei (Chap. 13 and Supplement 1.1) arriving in the upper Earth atmosphere. The atmosphere behaves as a hadronic calorimeter: the interactions of cosmic rays with atmospheric nuclei produce showers of hadrons. Among the hadrons, the neutrons, in particular, bombard the nitrogen nuclei, $^{14}_{7}$N, which are the major atmosphere constituent. This induces the reaction:

$$n + {}^{14}_{7}\mathrm{N} \rightarrow {}^{14}_{6}\mathrm{C} + p \tag{4.12}$$

Carbon has two stable isotopes: ^{12}C and ^{13}C. The $^{14}_{6}$C produced in the atmosphere has the (relatively) short half-life $t_{1/2} = 5730$ years. The amount of $^{14}_{6}$C in a sample is halved after 5730 years due to radioactive decay.

Due to the almost stationary cosmic ray flux on Earth, the production of $^{14}_{6}$C in the atmosphere is constantly occurring with a fixed rate since a very long time, so there is a fairly constant ratio of $^{14}_{6}$C to $^{12}_{6}$C atoms in the atmosphere. This ratio is approximately $(1.0 \div 1.3) \times 10^{-12}$.

When plants fix atmospheric carbon dioxide (CO_2) into organic material during photosynthesis, they incorporate a given quantity of $^{14}_{6}$C corresponding to the level of its concentration in the atmosphere. After their "death", plants are for instance used to make textiles or are consumed by other organisms (humans or other animals). Due to the metabolism of living organisms, also humans and animals have a $^{14}_{6}$C to $^{12}_{6}$C ratio at the level of the atmospheric concentration. From the instant of the vegetal or animal death, the $^{14}_{6}$C fraction in the organic material decreases due to its radioactive decay. The law of the radioactive decay is given in Sect. 4.5.2. Remember that in Eq. (4.42), $N(t) = N_0 e^{-t/\tau}$, the quantity τ is the lifetime which is related to the half-life, $t_{1/2}$, by the relation: $t_{1/2} = \tau \ln 2$.

After a time t, comparing the remaining $^{14}_{6}$C fraction of a sample to that expected from atmospheric $^{14}_{6}$C allows to measure the age of the sample. Its low activity limits the age determination by counting techniques to the order of 50,000 years. That can be extended to perhaps 100,000 years using more advanced accelerator techniques.

As the level of atmospheric ${}^{14}_{6}C$ is affected by variations in the cosmic ray intensity (which is in turn affected by variations in the Earth's magnetic field), high-accuracy measurements can only be achieved through a fine *calibration of raw, i.e., uncalibrated, radiocarbon dates*. The available standard calibration curves are based on the comparison of radiocarbon dates of samples that can be dated independently by other methods such as the examination of tree growth rings, deep ocean or ice sediment cores, lake sediments, coral samples, and others. The most accurate curve extends back quite accurately up to 26,000 years. Any errors in the calibration curve do not contribute more than ±16 y up to the last 6,000 y and no more than ±163 y over the entire 26,000 years [4R04].

Radiometric dating was extended to many other elements. For instance, the uranium-lead radiometric dating was used to date Earth rocks with a precision of less than two million years in four-and-a-half billion years (see Chap. 14).

Solutions

Problem 4.2 Consider, for example, the decay $e \to e\gamma$ and let see if, in the reference system in which the initial electron is at rest (the electron c.m. system), the energy and momentum conservation laws can be simultaneously satisfied. The momentum of the initial state is equal to zero. To have a momentum equal to zero in the final state, the photon must be emitted with momentum with the same magnitude and in the opposite direction with respect to the electron ($\mathbf{p}_e = -\mathbf{p}_\gamma$). But in this case, the final state energy is:

$$E_e + E_\gamma = \sqrt{m_e^2 c^4 + p_e^2 c^2} + |\mathbf{p}_\gamma| c$$

which is always larger than the initial energy $E_i = m_e c^2$, unless the photon energy is $E_\gamma = 0$ and therefore $p_e = 0$.

The radiation emission from a free electron is possible when (for instance) a recoiling nucleus is present, as shown in Fig. 4.2c.

Problem 4.4

(a) $d = \sqrt{(x_2 - x_1)^2 + (y_2 - y_1)^2 + (z_2 - z_1)^2} = 1.38$ m
(b) $t = 5.31 \cdot 10^{-9}$ s
(c) $v = \frac{d}{t} = \frac{1.38}{5.31 \cdot 10^{-9}} = 2.6 \cdot 10^8 \ \frac{\text{m}}{\text{s}}$
(d)

$$\beta = \frac{v}{c} = \frac{2.6 \cdot 10^8}{3 \cdot 10^8} = 0.87, \qquad \gamma = \frac{1}{\sqrt{1 - \beta^2}} = 2.03$$

$$t^* = \frac{t}{\gamma} = \frac{5.31 \cdot 10^{-9}\ \text{s}}{2.03} = 2.6\ \text{ns}$$

(e) The probability P that a particle with lifetime τ survives at the time t^* is $P = e^{-t^*/\tau}$. In this case: $\tau_{K^-} = 1.2 \cdot 10^{-8}$ s = 12 ns, and $P = e^{-2.6/12} = 0.80$. Then, only 20% of K^- of that velocity decay within a path length of 1.38 m.

Problem 4.5

(a) The radioactive decay law is $N(t) = N(0)e^{-t/\tau}$, where τ is the lifetime, $N(t)$ the number of nuclides surviving after a time t and $N(0)$ are those present at the initial time. In the case of the two carbon isotopes, since the $^{12}_{6}$C is stable, one has $N_{12}(t) = N_{12}(0)$ and:

$$\frac{N_{14}(t)}{N_{12}(t)} = \frac{N_{14}(0)e^{-t/\tau_{14}}}{N_{12}(0)} = 1.0 \cdot 10^{-12} e^{-t/\tau_{14}}$$

The half-life is related to the mean lifetime through the relation:

$$\frac{1}{2} = e^{-t_{14}^{1/2}/\tau_{14}} \longrightarrow \tau_{14} = t_{14}^{1/2}/\ln 2 = 8267 \text{ y}$$

The ratio between the numbers of nuclei of the two isotopes after a time $t^* =$ 12000 years, is therefore:

$$\frac{N_{14}(t^*)}{N_{12}(t^*)} = 1.0 \cdot 10^{-12} e^{-t^*/\tau_{14}} = 0.23 \cdot 10^{-12}$$

(b) In this case, the ratio between the two isotopes at the time $t = 0$ and the unknown time t^*, is 0.58; therefore, one must have:

$$\frac{N_{14}(t^*)}{N_{14}(0)} = \frac{N_{14}(0)e^{-t^*/\tau_{14}}}{N_{14}(0)} = 0.58 \longrightarrow t^* = -\tau_{14}\ln(0.58) = 4503 \text{ y}$$

Problem 4.6 The Rutherford differential cross-section is given in Eq. (4.54):

$$\left(\frac{d\sigma}{d\Omega}\right)_R = \frac{Z^2 z^2 e^4}{(4E_c)^2 \sin^4 \theta/2} \tag{4.13}$$

In this case, $z = 2$, $Z = 79$, $E_c = 8.0$ MeV. The solid angle $d\Omega = 2\pi \sin\theta d\theta$, inside which the scattering must be considered (indeed, there exists a symmetry in the azimuthal angle), is included in the range $[\theta_{min}, \theta_{max}]$ which is determined from the condition:

$$\tan\theta_{min} = 0.5/3.0 = 0.166 \longrightarrow \theta_{min} = 0.165 \text{ rad}$$

$$\tan\theta_{max} = 0.7/3.0 = 0.233 \longrightarrow \theta_{max} = 0.229 \text{ rad}$$

The cross-section σ is obtained by integrating Eq. (4.13):

$$\sigma = \int_{\theta_{min}}^{\theta_{max}} \left(\frac{d\sigma}{d\Omega}\right)_R 2\pi \sin\theta d\theta = \int_{\theta_{min}}^{\theta_{max}} \frac{Z^2 z^2 e^4}{(4E_c)^2} \frac{2\pi \; 2\sin\frac{\theta}{2}\cos\frac{\theta}{2}}{\sin^4\frac{\theta}{2}} 2d\frac{\theta}{2}$$
$$\sigma = \frac{Z^2 z^2 e^4}{(4E_c)^2} \frac{2\pi}{\sin^2\frac{\theta}{2}}\Bigg|_{\theta_{max}}^{\theta_{min}} \tag{4.14}$$

It is clear that for $\theta_{min} \to 0$, Eq. (4.14) is divergent: this is due to the fact that the Rutherford formula is meaningless at small angles since the finite dimensions of the target and projectile were neglected.

Equation (4.14) is expressed in the c.g.s. unit system, in which $e = 4.8 \cdot 10^{-10}$ esu, 8 MeV $= 8 \times 1.6 \cdot 10^{-6}$ erg. Inserting the numerical values, one obtains:

$$\sigma = \frac{79^2 \cdot 2^2 (4.8 \cdot 10^{-10})^4}{4^2 (1.24 \cdot 10^{-5})^2} \frac{2\pi}{\sin^2\frac{\theta}{2}}\Bigg|_{\theta_{max}}^{\theta_{min}} = 5.0 \times 10^{-25}\, 2\pi[37.1 - 19.4]$$
$$= 5.7 \times 10^{-23} \text{ cm}^2 \tag{4.15}$$

for a scattering angle within the range $[\theta_{min}, \theta_{max}]$. To obtain the number of scattered particles, one must calculate the number of gold nuclei per area unit of a sheet 4.0 μm thick. The atomic number of gold is $A = 197$ and the density is $\rho = 19.3$ g/cm^3; therefore, a mole of gold has a mass $M_A = 197$ g and the number of atoms per cm^3 is:

$$N = \frac{\rho N_A}{M_A} = 5.87\; 10^{22} \text{ atoms/cm}^3$$

The number of atoms per area unit of a sheet $d = 4 \times 10^{-4}$ cm thick, is therefore:

$$n = Nd = 5.87 \cdot 10^{22} \times 4 \cdot 10^{-4} = 2.35 \cdot 10^{19} \text{ cm}^{-2}$$

The number of particles that enter the detector per second is thus:

$$N_{det} = \Phi n \sigma = \left(5.0 \cdot 10^7\right)\left(2.35 \cdot 10^{19}\right)\left(5.7 \times 10^{-23}\right) = 6.7 \cdot 10^4 \text{ s}^{-1}$$

which corresponds to 0.13% of the initial beam.

Problem 4.7 Here, natural units ($\hbar = c = 1$) are used. For the electromagnetic interaction, the matrix element $|M_{if}|$ can be replaced with the bosonic propagator given in Eq. (4.32), $f(q) = \frac{g_0 g}{q^2 + m^2}$, with $m = 0$ since the photon is massless, and $g/4\pi = e$, $g_0 = e$; thus, Eq. (4.28) becomes:

$$W = 2\pi \left(\frac{4\pi e^2 \psi(0)}{q^2}\right)^2 \frac{dN}{dE_0} \tag{4.16}$$

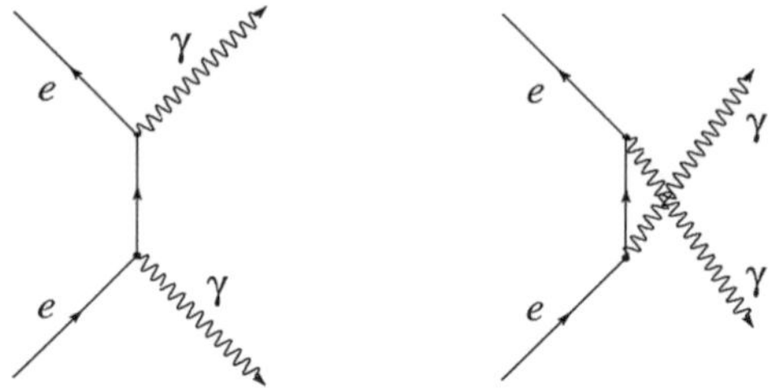

Fig. 4.1 Feynman diagrams for the annihilation $e^+e^- \rightarrow \gamma\gamma$

The term $\psi(0)$ must be inserted to take into account that the electron and positron are not free particles (described by a wave function of the type given in Eq. (4.11)), but bound in a particular state (i.e., positronium) with a generic wave function $\psi(x)$. The quantity $|\psi(0)|^2$ determines the probability for the e^+, e^- to be in the same region of space.

The phase space factor in (4.16) is that of a two-body decay, as discussed in Sect. 8.6.1; it is given by: $dN/dE_0 = E_0^2/2\pi^2$. In addition, from the Feynman diagrams (see Fig. 4.1), one can deduce that, for the annihilation process in the c.m. system, one has $s = m_e^2 = E_0^2 = q^2$. Since in the natural unit system, one has $\alpha = e^2$, one can finally rewrite (4.16) as

$$W = \Gamma\left({}^1S_0 \rightarrow \gamma\gamma\right) = 2\pi\left(\frac{16\pi^2\alpha^2|\psi(0)|^2}{m_e^4}\right)\frac{m_e^2}{2\pi^2} = \frac{16\pi\alpha^2}{m_e^2}\left|\psi(0)\right|^2 \tag{4.17}$$

Warning: note that we have indicated $W = \Gamma$. In our textbook, W has the dimension of $[time]^{-1}$ (Sect. 4.5), while Γ has the dimension of $[energy]$ (Sect. 7.5). However, remember that W represents the inverse of the lifetime: $W = \tau^{-1}$ and that the lifetime and the total width Γ are related through the uncertainty principle (Sect. 7.4.2): $\tau\Gamma = \Gamma/W = \hbar$. Therefore, in the natural unit system in which $\hbar = 1$, one has $\Gamma = W$.

In Eq. (4.17), a factor of 4 is missing because the spins were not taken into account so far. In the initial state, the spin is zero. The spin of the photon is $s_\gamma = 1$. The spin multiplicity of the final state is not $(2s_\gamma + 1)^2 = 9$, but 4 because the photon has only two polarization states. The conservation of the angular momentum allows only 2 of the 4 states. Another factor of two arises from the fact that the two photons are indistinguishable (generating the two diagrams of Fig. 4.1). Finally, the total width of the process is:

$$\Gamma\left({}^1S_0 \rightarrow \gamma\gamma\right) = \frac{4\pi\alpha^2}{m_e^2}\left|\psi(0)\right|^2 \tag{4.18}$$

Problem 4.8 In the Bohr theory, the positronium would never annihilate. In fact, Bohr theory requires stationary orbits with $m_e vr = \hbar$ and the positron and electron can not overlap. Quantum mechanics provides orbits described by a probability density function. Equation (4.2) has a finite value at the origin, $r = 0$. The positronium wave function is derived from that of the hydrogen atom replacing the value of the Bohr radius with that of the positronium. The Bohr radius is obtained assuming an infinite mass for the proton. For the positronium, the reduced mass of the two electron system must be taken into account. This implies that in Eq. (4.3), one must replace $m_e \to \mu_e = m_e/2$. The positronium radius a' has a value twice that of the hydrogen atom: $a' = 2a$. The wave function at the origin is therefore: $\psi(0) = \frac{1}{\sqrt{\pi}(2a)^{3/2}}$. Finally, Eq. (4.18) becomes:

$$\Gamma(^1S_0 \to \gamma\gamma) = \frac{4\pi\alpha^2}{m_e^2}\frac{1}{\pi(2a)^3} = \frac{4\pi\alpha^2}{m_e^2}\frac{(m_e\alpha)^3}{8\pi} = \frac{m_e\alpha^5}{2} = 5.2 \cdot 10^{-12}\ \text{MeV} \tag{4.19}$$

The lifetime τ is related to the width Γ by the uncertainty principle:

$$\tau = \frac{\hbar}{\Gamma} = \frac{6.6 \cdot 10^{-22}}{5.2 \cdot 10^{-12}} = 1.3 \cdot 10^{-10}\ \text{s}$$

The measured value [4K04] is $\tau_{mea} = 1.244 \cdot 10^{-10}$ s.

Problem 4.9 As we saw in Appendix 4 and Eq. (A.17), the most simple Hamiltonian operator for a free particle has the form: $H = \boldsymbol{\alpha} \cdot \mathbf{p} + \beta m$, where $\boldsymbol{\alpha}$ and β are the matrices defined in (A.19). The equation $E\psi = i\hbar\frac{\partial}{\partial t}\psi = H\psi$ can therefore be written as:

$$i\hbar\frac{\partial}{\partial t}\psi + i\hbar c\boldsymbol{\alpha} \cdot \nabla\psi - \beta mc^2\psi = 0.$$

Multiplying on the left by β, using natural units $c = 1$ and defining:

$$\gamma = \beta\boldsymbol{\alpha}, \qquad \gamma^i = \beta\alpha_i \quad i = 1, 2, 3, \qquad \gamma^0 = \beta$$

the following equation is obtained:

$$i\hbar\gamma^0\frac{\partial}{\partial t}\psi + i\hbar\boldsymbol{\gamma} \cdot \nabla\psi - m\psi = 0.$$

Problem 4.10 In the four-spinor representation, the $\boldsymbol{\sigma}$ Pauli generalized matrix in Eq. (4.7) is

$$\boldsymbol{\sigma} = \begin{vmatrix} \boldsymbol{\sigma} & 0 \\ 0 & \boldsymbol{\sigma} \end{vmatrix}. \tag{4.20}$$

The total angular momentum $\mathbf{J}$ commutes with the Hamiltonian H if:

$$\left[\left(\hbar\mathbf{L} + \frac{1}{2}\hbar\boldsymbol{\sigma}\right), H\right] = 0. \tag{4.21}$$

This correspond to the condition that both $[\mathbf{L}, H] = 0$ and $[\boldsymbol{\sigma}, H] = 0$. By definition, it is:

$$\mathbf{L} = L_x\mathbf{i} + L_y\mathbf{j} + L_z\mathbf{k} = \mathbf{r} \times \mathbf{p} = (yp_z - zp_y)\mathbf{i} + (zp_x - xp_z)\mathbf{j} + (xp_y - yp_x)\mathbf{k} \tag{4.22}$$

The condition $[\mathbf{L}, H] = 0$ is demonstrated using the following commutation rules:

$$[x_j, x_k] = [p_j, p_k] = 0, \qquad [x_j, p_k] = i\hbar\delta_{ij} \tag{4.23a}$$

(the second condition comes from the quantum definition of momentum, $\mathbf{p} = -i\hbar\nabla$) and

$$[L_j, \beta] = 0, \qquad [\boldsymbol{\alpha}, \mathbf{r}] = [\boldsymbol{\alpha}, \mathbf{p}] = 0. \tag{4.23b}$$

Since the α and β matrices do not depend neither on $\mathbf{r}$ nor on $\mathbf{p}$, one has:

$$\begin{aligned}
[L_x, H] &= [L_x H - H L_x] = i\hbar(\alpha_y p_z - \alpha_z p_y)\\
[L_y, H] &= [L_y H - H L_y] = i\hbar(\alpha_z p_x - \alpha_x p_z)\\
[L_z, H] &= [L_z H - H L_z] = i\hbar(\alpha_x p_y - \alpha_y p_x)
\end{aligned}$$

Analogously, the condition $[\boldsymbol{\sigma}, H] = 0$ follows from the commutation rules:

$$[\sigma_k, \alpha_k] = [\sigma_k, \beta] = 0, \qquad [\sigma_j, \alpha_k] = 2i\alpha_k \begin{pmatrix} 0 & 1 \\ 1 & 0 \end{pmatrix} \tag{4.24}$$

from which follows that:

$$\begin{aligned}
\left[\frac{\hbar}{2}\sigma_x, H\right] &= \frac{\hbar}{2}[\sigma_x H - H\sigma_x] = i\hbar c(\alpha_y p_z - \alpha_z p_y)\\
\left[\frac{\hbar}{2}\sigma_y, H\right] &= \frac{\hbar}{2}[\sigma_y H - H\sigma_y] = i\hbar c(\alpha_z p_x - \alpha_x p_z)\\
\left[\frac{\hbar}{2}\sigma_z, H\right] &= \frac{\hbar}{2}[\sigma_z H - H\sigma_z] = i\hbar c(\alpha_x p_y - \alpha_y p_x)
\end{aligned}$$

and finally $[(\hbar\mathbf{L} + \frac{1}{2}\hbar\boldsymbol{\sigma}),\ H] = 0$.

Problem 4.11 It is easy to demonstrate that:

$$\boldsymbol{\alpha}\cdot\mathbf{p} = \begin{pmatrix} 0 & \boldsymbol{\sigma} \\ \boldsymbol{\sigma} & 0 \end{pmatrix}\cdot\mathbf{p} = \boldsymbol{\alpha}\cdot\mathbf{p}\begin{pmatrix} 0 & 1 \\ 1 & 0 \end{pmatrix} = \begin{pmatrix} 0 & 1 \\ 1 & 0 \end{pmatrix}\boldsymbol{\alpha}\cdot\mathbf{p} \tag{4.25}$$

The Hamiltonian operator can be written as:

$$H = \begin{pmatrix} 0 & 1 \\ 1 & 0 \end{pmatrix}\boldsymbol{\alpha}\cdot\mathbf{p} + \begin{pmatrix} 1 & 0 \\ 0 & -1 \end{pmatrix} m. \tag{4.26}$$

From the definition of Λ and from (4.25) and (4.26), it follows:

$$[\Lambda, H] = [\Lambda H - H\Lambda] = 0. \tag{4.27}$$

Problem 4.12 Let us insert the wave function expression

$$\psi = \begin{pmatrix} \phi \\ \chi \end{pmatrix} e^{i(\mathbf{p}\cdot\mathbf{r}-Et)}$$

in the compact form of the Dirac equation:

$$i\sum_{\mu} \gamma^{\mu} \frac{\partial}{\partial x_{\mu}} \psi - m\psi = 0$$

where $\mu = 0, \ldots, 3$.

Performing the derivatives and taking into account the matrices α_i and β defined in (A.19), the following matrix equation is obtained:

$$(E - \boldsymbol{\alpha}\cdot\mathbf{p} - \beta m)\begin{pmatrix} \phi \\ \chi \end{pmatrix} = 0.$$

which corresponds to:

$$\left[\begin{pmatrix} 1 & 0 \\ 0 & 1 \end{pmatrix} E - \begin{pmatrix} 0 & \boldsymbol{\sigma} \\ \boldsymbol{\sigma} & 0 \end{pmatrix} \mathbf{p} - \begin{pmatrix} 1 & 0 \\ 0 & -1 \end{pmatrix} m\right]\begin{pmatrix} \phi \\ \chi \end{pmatrix} = \begin{pmatrix} 0 \\ 0 \end{pmatrix}$$

from which one obtains:

$$E\phi - \boldsymbol{\sigma}\cdot\mathbf{p}\chi - m\phi = 0$$

$$E\chi - \boldsymbol{\sigma}\cdot\mathbf{p}\phi + m\chi = 0$$

or

$$\phi = \frac{\boldsymbol{\sigma}\cdot\mathbf{p}}{E-m}\chi, \qquad \chi = \frac{\boldsymbol{\sigma}\cdot\mathbf{p}}{E+m}\phi.$$

References

[4B90] Bowman, S.: Radiocarbon Dating (Interpreting the Past). University of California Press, Berkeley (2000). ISBN: 978-0520070370

[4K04] Karshenboim, S.G.: Precision study of positronium: testing bound state QED theory. Int. J. Mod. Phys. A **19**, 3879 (2004). arXiv:hep-ph/0310099 [hep-ph]

[4R04] Reimer Paula, J., et al.: INTCAL04 terrestrial radiocarbon age calibration, 0-26 Cal Kyr BP. Radiocarbon **46**(3), 1029–1058 (2004)

Chapter 5
First Discussion of the Other Fundamental Interactions

Problems

5.1. **Neutron and antineutron**. How do you distinguish a neutron from an antineutron? And how can you distinguish a neutrino from the corresponding antineutrino?
[See solutions]

5.2. **Gravity and electric forces**. Estimate the size of a hypothetical hydrogen atom whose electron and proton are bound only by the gravitational force.
[See solutions]

5.3. **Feynman diagrams**. Draw the Feynman diagrams that illustrate the following reactions; indicate the involved fundamental interaction.
(a) $e^- p \to e^- p$
(b) $e^+ e^- \to \nu_e \bar{\nu}_e$
(c) $e^- p \to \nu_e n$
(d) $ud \to udd\bar{d}$
[See solutions]

5.4. **Tau decay**. Consider the decay $\tau^- \to e^- + X$.
(a) Which neutral particles form the system X?
(b) Draw the Feynman diagram for this process.
[See solutions]

5.5. **Δ^{++} decay**. Estimate the decay fraction ratio

$$\Gamma(\Delta^{++} \to p e^+ \nu_e)/\Gamma(\Delta^{++} \to p\pi^+)$$

[See solutions]

S. Braibant et al., *Particles and Fundamental Interactions: Supplements, Problems and Solutions*, Undergraduate Lecture Notes in Physics,
DOI 10.1007/978-94-007-4135-5_5, © Springer Science+Business Media Dordrecht 2012

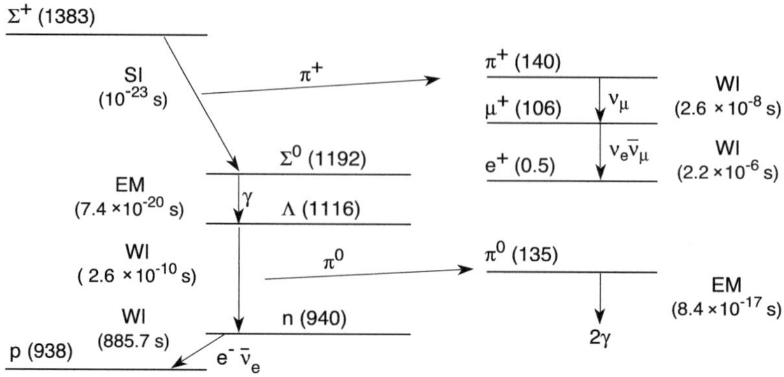

Fig. 5.1 Decay chain of the $\Sigma^+(1383)$ baryon. The lifetime of each decay channel are reported in parentheses

5.6. **Neutron star**. A neutron star is produced by the gravitational collapse of stars more massive than $8M_\odot$, where $M_\odot$ is the mass of the Sun. In the center of the star remains a sort of giant nucleus, consisting of only neutrons, held together by its own gravitational force. Calculate the radius of a neutron star assuming that:
   (i) the mass of the nucleus is $1.4M_\odot$ and
   (ii) the density of the neutron star is constant.
   [See solutions]

5.7. $\boldsymbol{\Sigma^+}$ **decay**. Explain the decay chain of the $\Sigma^+(1383)$ baryon, see Fig. 5.1. In terms of quarks, the Σ^+ is made of quarks $[uus]$.
   [See solutions]

Supplement 5.1: Baryon Number Conservation: the Search for Proton Decay

In the Standard Model of particle physics (see Chap. 11), quarks and leptons are placed in separate multiplets and the baryon number conservation forbids the proton decay. However, there is no known gauge symmetry which generates baryon number conservation. Therefore, the validity of baryon number conservation must be considered as an experimental question. Grand Unified Theories (GUTs, see Chap. 13) place quarks and leptons in the same multiplets; therefore, quark $\leftrightarrow$ lepton transitions are possible: they are mediated by massive X, Y bosons, respectively with electric charges of $4/3$ and $1/3$ (see Fig. 13.3). Because the mass of these gauge bosons are predicted to be very large, such processes are expected to be very rare.

Starting from the 1980s, the search for proton decay was the main reason for developing underground laboratories and large detectors [5P84]. The simplest GUT model, SU(5), predicts a proton lifetime value of $\tau_p \sim 10^{30}$ years for the process $p \to e^+\pi^0$. The expected number of events per year in a decay channel with branching ratio BR is

$$N = (f \cdot N_N \cdot M_F \cdot T \cdot BR \cdot \epsilon)/\tau_p \tag{5.1}$$

where f is the proton fraction ($f = p/(p+n)$) in the target, N_N is the number of nucleons in 1 kt mass ($\sim 6 \cdot 10^{23} \cdot 10^9$), M_F (in kt) is the fiducial mass (an olympic swimming pool contains ~ 3 kt of water), T is the detector lifetime (in years) and ϵ is the overall detection efficiency. The simplest GUT prediction leads to the possible observation of many proton decay events in a kt-scale detector.

For nucleon decay searches, the most competitive experimental techniques use water Cherenkov detectors and tracking calorimeters. The pioneering proton decay experiments were water Cherenkov (IMB, Kamiokande) and tracking calorimeters (KGF, NUSEX, Soudan). The latter consists of sandwiches of iron plates and ionization/scintillation detectors. The Cherenkov detection allows for larger masses, while tracking calorimeters provide better space resolution and good identification of electrons, muons and charged kaons. The main background comes from low energy atmospheric neutrinos. A large fraction of this background can be eliminated by selecting events well contained in the detector fiducial volume and by applying topological and kinematical constraints. The largest experiment currently taking data is Super-Kamiokande (SK), with a total mass of ~50 kt.

The proton decay $p \to e^+\pi^0$ ($\mu^+\pi^0$) is almost background free. The proton at rest decays in two light particles, carrying the ~ 1000 MeV of its rest mass. Let us shortly describe the procedure used by SK [5K09] to set the strongest limits on the proton lifetime ($>8.2 \times 10^{33}$ and 6.6×10^{33} years at the 90% confidence level for $p \to e^+\pi^0$ and $p \to \mu^+\pi^0$, respectively). To reduce the contamination from external neutrons, the detector volume was restricted to the inner 22.5 kt (*fiducial volume*) of the detector which is surrounded by 11,146 20-inch diameter photomultiplier tubes (PMTs). The e^+ (or μ^+) produces Cherenkov light. The photons arising from the π^0 decay, $\pi^0 \to \gamma\gamma$, originate an electromagnetic shower which produces a signal similar to that of the electron. The intersection of the Cherenkov light cone with the instrumented detector surface generates signals having the shape of "rings". In multi-ring events, as expected for hypothetic proton decays, the directions and opening angles of visible Cherenkov rings were reconstructed, and each identified ring was classified as a showering electron-like ring or a non-showering muon-like ring.

The particle momentum was estimated by the number of detected photoelectrons. Kinematics constraints were used. The total momentum in an event is $P_{tot} = |\sum_i \mathbf{p}_i|$, where $\mathbf{p}_i$ is the reconstructed momentum vector of the i-th ring. The total invariant mass is $M_{tot} = \sqrt{E_{tot}^2 - P_{tot}^2}$, where the total energy $E_{tot} = \sum_i \sqrt{p_i^2 + m_i^2}$, and m_i is assumed to be the electron or muon rest mass, according to the particle identification through ring shapes. In order to select proton decay signals, the following selection criteria were applied:

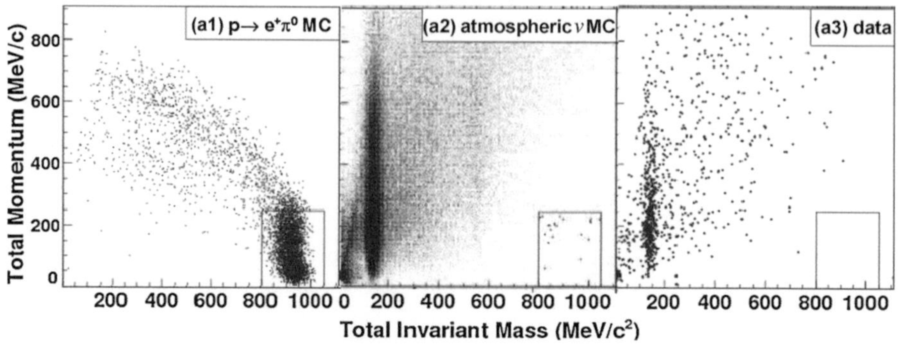

Fig. 5.2 Total momentum versus total invariant mass distributions for Monte Carlo (MC) proton decay events (*left panel*), 500 year-equivalent MC atmospheric neutrino events (*central panel*) and data (*right panel*). All events satisfy the selection criteria for $p \to e^+\pi^0$ except for (E). The *boxes* in the plots indicate the selection criterion described in (E). *Points* in the *signal box* of the atmospheric neutrino MC are shown with a larger size

(A) the number of rings is two or three; (B) one of the rings is e-like (μ-like) for $p \to e^+\pi^0$ ($p \to \mu^+\pi^0$) and all the other rings are e-like; (C) for three ring events, the reconstructed π^0 invariant mass is between 85 and 185 MeV/c^2; (D) the muon can eventually decay in the detector; (E) the reconstructed total momentum is smaller than 250 MeV/c, and the reconstructed total invariant mass is between 800 and 1050 MeV/c^2.

The proton decay signal sample and the background sample from atmospheric neutrino interactions were simulated using Monte Carlo procedures, which take into account accurate simulation of the detector response. For each simulated signal and background event, the reconstructed total momentum and total energy were inserted in a two dimensional plot (first two panels of Fig. 5.2). From the null signal (no real data are present in the box signal of the third panel of Fig. 5.2), using Eq. (5.1), SK was able to derive the limits quoted above. These experimental limits exclude without any doubt the simplest GUT models (which predict a proton decay lifetime smaller than the experimental lower limits). Many more proton decay modes were searched for, all with null results. At present, larger project are proposed to improve by more than one order of magnitude the limit on the proton lifetime (Hyper-Kamiokande and LAGUNA projects).

Underground laboratories (mainly motivated by the need to reduce the background in proton decay experiments) are also used for different experimental studies:

- Detection of low energy phenomena, $E < 20$ MeV, as solar neutrinos, neutrinos from stellar gravitational collapse, the search for neutrino-less double-beta decays and searches for rare phenomena. The main problem is the radioactivity background; refined detectors, often of large mass, are needed. For the detection of low energy neutrinos, the most important parameters are the detector mass and the energy threshold ($\sim$ MeV);

- Study of ~1 GeV events, like in nucleon decays and in atmospheric/beam neutrino oscillations. The main feature of a detector is its mass (1–50 kt) and the capability of identifying neutrino events;
- Detection of through-going particles, high energy muons, monopole candidates, etc. The main feature of these detectors is the area.

Solutions

Problem 5.1 The neutron decays into $n \to pe^-\overline{\nu}_e$: the lighter charged particle in the presence of a known magnetic field behaves as a negatively charged particle. The antineutron decays into $\overline{n} \to \overline{p}e^+\nu_e$: the lighter charged particle in the presence of a known magnetic field behaves as a positive particle.

When the neutrino interacts (excluding neutral current interactions), it produces a negatively charged lepton (the ν_e produces an electron, the ν_μ a muon). The antineutrino always produces an antilepton, with a positive electric charge.

Problem 5.2 The module of the gravitational force F_G and of the Coulomb force F_E acting between the proton and the electron in the hydrogen atom are respectively:

$$F_G = G_N \frac{m_p m_e}{r^2}; \qquad F_E = \frac{1}{4\pi\epsilon_0}\frac{e^2}{r^2}.$$

Inserting the numerical values of m_p, m_e, the gravitational constant G_N and the vacuum dielectric constant ϵ_0, their ratio is:

$$\frac{F_G}{F_E} = 4\pi\epsilon_0 G_N \frac{m_p m_e}{e^2} = 5 \times 10^{-40}.$$

The radius of an atom bounded by the gravitational force can be determined using the angular momentum quantization rule and the orbit stability condition as in the Bohr atom:

$$\begin{cases} m_e v r = \hbar \\ m_e v^2 r^{-1} = F. \end{cases}$$

Obtaining $m_e v = \hbar/r$ from the first equation and substituting the corresponding quantity in the second equation, one finds $r^3 = \hbar^2/F \cdot m_e$; therefore, the radius r is:

$$r = \left(\frac{\hbar^2}{F \cdot m_e}\right)^{1/3}.$$

The ratio between the Bohr radius of the hydrogen atom $a = 0.5 \times 10^{-10}$ m where $F = F_E$ and the radius r_G of an atom bounded by the gravitational force ($F = F_G$) is:

$$\frac{a}{r_G} = \left(\frac{F_G}{F_E}\right)^{1/3} = (5 \times 10^{-40})^{1/3} = 8 \times 10^{-14}$$

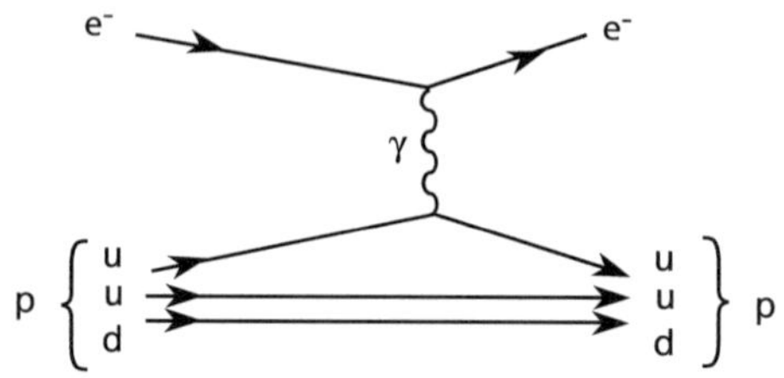

Fig. 5.3 $e^- p \to e^- p$ elastic scattering

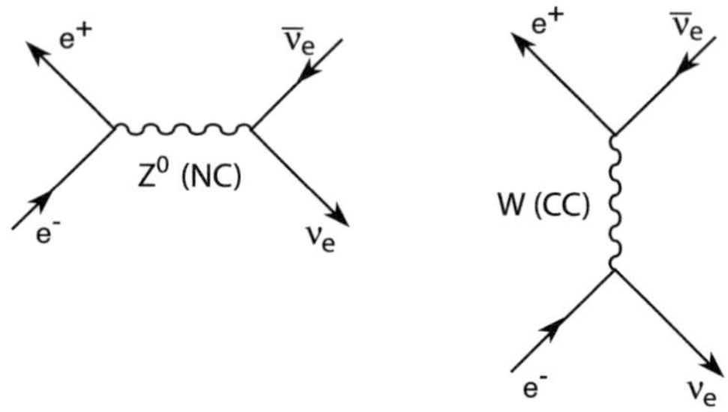

Fig. 5.4 $e^+e^- \to \nu_e\bar{\nu}_e$ annihilation: (*left*) weak NC interaction through a Z^0 boson exchange and (*right*) weak CC interaction through a $W^\pm$ boson exchange

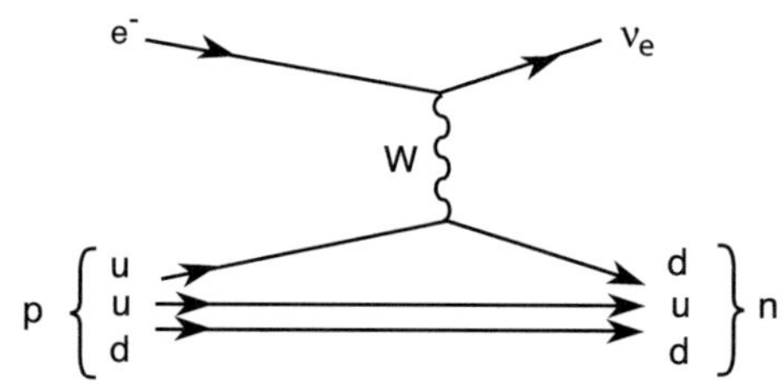

Fig. 5.5 $e^- p \to \nu_e n$ CC interaction

from which $r_G = (0.5 \times 10^{-10})/(5 \times 10^{-14}) = 625$ m.

Problem 5.3

(a) $e^- p \to e^- p$. The fundamental force involved is the electromagnetic interaction. A negligible contribution from the weak interaction (when the photon in the figure is replaced by a Z^0) can also be assumed (Fig. 5.3).

(b) $e^+e^- \to \nu_e\bar{\nu}_e$. The fundamental force involved is the weak interaction. The weak interaction (Chap. 8) can take place through the exchange of a charged massive vector boson, the $W^\pm$, or through a neutral one, the Z^0. The exchange of a $W^\pm$ corresponds to a weak *charged current interaction* (CC), while the exchange of a Z^0 is a weak *neutral current interaction* (NC). The electron-positron annihilation occurs both in NC and in CC, as shown in Fig. 5.4.

(c) $e^- p \to \nu_e n$. The fundamental force involved is the weak charged current interaction, involving a u quark of the proton (the others two quarks act as spectators) and an electron (Fig. 5.5).

(d) $ud \to udd\bar{d}$. The fundamental force involved is the strong interaction with the radiation and successive decay of a gluon.

Fig. 5.6 $ud \to udd\bar{d}$

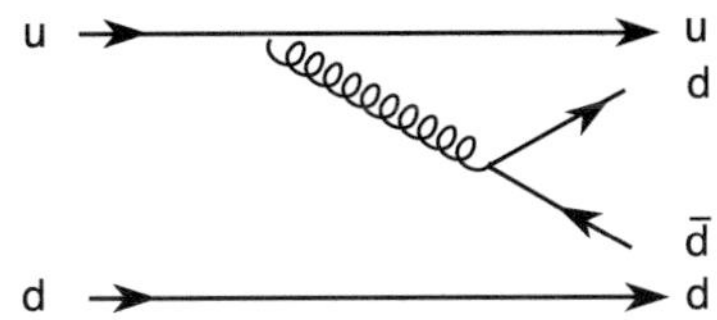

Free quarks have not been observed so far. This process always occurs inside hadrons, with one (or more) spectator quark. For instance, if the presence of an additional spectator u quark is assumed, the process shown in Fig. 5.6 can be that of a spin 3/2 $\Delta^+ = [uud]$ which decays into a spin 1/2 proton ($= [uud]$) plus a $\pi^0 = [d\bar{d}]$.

Problem 5.4

(a) In the reaction, besides the electric charge, the lepton family numbers L_e, L_ν, L_τ must be preserved. The status $[X]$ must therefore have the following quantum numbers:

$$[X] = [L_\tau = 1;\ L_e = -1;\ Q = 0].$$

In the final state, one particle of the doublet $\binom{\nu_\tau}{\tau^-}$ and one of the doublet $\binom{\overline{\nu}_e}{e^+}$ must thus be present. From the first doublet, the particle cannot be a τ^- because the energy conservation in the rest frame of the decaying τ would be violated. It must therefore be a ν_τ. To ensure the electric charge conservation, the positron cannot be present and the member of the second doublet that must be used, is the $\overline{\nu}_e$.
Therefore, $X = \nu_\tau \overline{\nu}_e$.

(b) The Feynman diagram for the τ decay is shown in Problem 8.3b.

Problem 5.5 The $\Delta^{++} \to p e^+ \nu_e$ decay is due to the weak interaction, with a u-type quark that transforms into a d-type quark by emitting a W^+. The vertices connected by the W vector boson are characterized by a coupling constant $\alpha_W \sim 10^{-5}$.

The $\Delta^{++} \to p\pi^+$ decay is due to the strong interaction, with a coupling constant $\alpha_s \sim 1$. The expected ratio between branching ratios is then:

$$\frac{\Gamma(\Delta^{++} \to p e^+ \nu_e)}{\Gamma(\Delta^{++} \to p\pi^+)} \simeq \left(\frac{10^{-5}}{1}\right)^2 \sim 10^{-10}.$$

The weak interaction channel is so rare that it has never been observed so far.

Problem 5.6 Nuclei are incompressible and the volume of a nucleus corresponds to the volume of its constituents. A proton has mass $m_p = 1.6 \times 10^{-24}$ g and its volume is that of a sphere with radius $R_0 \sim 1.3 \times 10^{-13}$ cm. The density of matter inside the proton radius (the nuclear density) is:

$$\rho_N = \frac{m_p}{\frac{4\pi}{3} R_0^3} = 2 \times 10^{14} \text{ g/cm}^3$$

The sun mass is $M_\odot = 2 \times 10^{33}$ g. The considered neutron star mass is $M_{NS} = 1.4 M_\odot = 2.8 \times 10^{33}$ g. Assuming that the neutron star density be constant, one has $\rho_{NS} = \rho_N = M_{NS}/(4\pi/3R^3)$. It follows that the radius R is:

$$R = \sqrt[3]{\frac{3M_{NS}}{4\pi\rho_n}} = \sqrt[3]{\frac{3 \cdot 2.8 \cdot 10^{33}}{4\pi 2 \cdot 10^{14}}} = 1.45 \times 10^6 \text{ cm} = 14.5 \text{ km}$$

Problem 5.7 The decays presented here are briefly analyzed; they will be discussed in detail in later chapters. The quark composition of the particles is given in the squared brackets. The first decay is due to the strong interaction (process similar to that shown in Fig. 5.6):

$$\Sigma^+ = [usu] \to us(d\overline{d})u \to [usd][\overline{d}u] = \Sigma^0\pi^+$$

In this case, it was energetically favorable to produce a $(d\overline{d})$ quark-antiquark pair. The quarks are then combined to form the two physical states $\Sigma^0\pi^+$. The lifetime is that typical of strong interaction decays ($\sim 10^{-23}$ s).

The Σ^0 has a composition $[uds]$, the same as that of the Λ^0 baryon. However, the Σ^0 is part of a particle triplet (with Σ^+, Σ^-) with similar characteristics. The strong isospin value $I = 1$ is attributed to the Σ triplet. The Λ^0 is instead a singlet state with $I = 0$. This quantum number represents a symmetry of the wave function which is conserved by the strong interaction, but not by the electromagnetic interaction. The $\Sigma^0 \to \Lambda^0\gamma$ decay is indeed due to the electromagnetic interaction, with a characteristic lifetime of 10^{-19} s.

The Λ^0 decays through the weak interaction (see Fig. 8.17b), with a lifetime of $\sim 10^{-10}$ s:

$$\Lambda^0 = [uds] = (ud)s \to (ud)W^-u \to (ud)(d\overline{u})u \to [udd][\overline{u}u] = n\pi^0$$

The s quark can change into a different quark flavor only via the weak interaction. The W^- vector boson decays in turn into a $d\overline{u}$ pair.

Finally, the neutron decays in $pe^-\overline{\nu}_e$ through the weak interaction, with a lifetime ~900 s. Note that the last two processes have long lifetimes compared to those typical of the electromagnetic and strong interactions. However, they differ from each other by about 12 orders of magnitude. An additional important factor in the decay is the free energy E_0 available in the final state: the transition probability depends on E_0^5 in the case of a three body decay. For the Λ^0, the available energy is $E_0^\Lambda = m_\Lambda - m_n \simeq 175$ MeV; in the case of the neutron, one has $E_0^n \simeq 1$ MeV. Finally, as shown in Chap. 8, the decay of a strange quark $s \to uW^-$ is suppressed by a factor of 20 compared to the decay of a d quark $d \to uW^-$. The product of these two factors provides the difference between the Λ^0 and n lifetimes.

References

[5P84] Perkins, D.H.: Proton decay experiments. Annu. Rev. Nucl. Part. Sci. **34**, 1–50 (1984)
[5K09] Nishino, H., et al. (Super-K Collaboration): Search for proton decay via $p \to e^+\pi^0$ and $p \to \mu^+\pi^0$ in a large water Cherenkov detector. Phys. Rev. Lett. **102**, 141801 (2009)

Chapter 6
Invariance and Conservation Principles

Problems

6.1. **Electric dipole of the neutron**. Show that if a particle (for instance, the neutron) has a non vanishing electric dipole moment $\mathbf{d}_E$, both parity and time reversal invariance are violated.
[See solutions]

6.2. **CP of neutron decay**. Consider the β neutron decay, $n \to p + e^- + \overline{\nu}_e$, and apply the parity operator; does the resulting process exist in nature? Then, apply the charge conjugation operator. What kind of process do you obtain? What can be concluded about the CP operator in β decay?
[See solutions]

6.3. **η^0 decay**. **Pseudoscalar meson**. The pseudoscalar meson η^0 has quantum numbers $J^{PC} = 0^{-+}$, mass 548 MeV, full width $\Gamma = 1.30$ keV. The main decay modes (from the PDG) and corresponding *branching ratios* (BR) are:

Decay	BR
$\eta^0 \to \gamma\gamma$	0.393
$\eta^0 \to \pi^0\pi^0\pi^0$	0.326
$\eta^0 \to \pi^0\pi^+\pi^-$	0.227

There are also upper limits for the following decays:

Decay	BR
$\eta^0 \to \gamma\gamma\gamma$	$<1.6 \cdot 10^{-5}$
$\eta^0 \to \pi^0\pi^0$	$<3.5 \cdot 10^{-4}$
$\eta^0 \to \pi^0\gamma$	$<0.9 \cdot 10^{-6}$
$\eta^0 \to e^+\mu^-$ or $e^-\mu^+$	$<6 \cdot 10^{-6}$

S. Braibant et al., *Particles and Fundamental Interactions: Supplements, Problems and Solutions*, Undergraduate Lecture Notes in Physics,
DOI 10.1007/978-94-007-4135-5_6, © Springer Science+Business Media Dordrecht 2012

(a) Determine the lifetime of the η^0 decay and which interaction induces the decay.
(b) Explain, using conservation laws, the reason why the decays with the upper limits quoted above were not observed. Remember that the photon has $J^{PC}=1^{--}$, and the π^0 has $J^{PC}=0^{-+}$.
[See solutions]

6.4. ***Σ* decay.** Explain why the Σ^0 hyperon decays into $\Lambda^0\gamma$ instead of $n\pi^0$.
[See solutions]

6.5. **Ω^- decay.** Discuss the possible decay modes of the Ω^- particle allowed by conservation laws. Show that the weak interaction decay is the only possible one.
[See solutions]

6.6. **Pion decay.** The π^0 meson has spin zero and mass $m_\pi = 135$ MeV/c^2, and decays into two photons. Since the measurement of the characteristics of the final state photons provides information on the π^0 spin/parity, determine:
(a) the angular distribution of the emitted photons in the π^0 rest frame;
(b) the shape of the energy spectrum of the emitted photons in the laboratory system;
(c) the maximum and minimum energy of the emitted photons when the π^0 has an energy of 0.8 GeV.
[See solutions]

6.7. **Conservation law rules.** Verify if the following reactions satisfy all conservation laws:
(a) $\overline{K^0}p \to K^-p\pi^+$
(b) $\pi^-p \to K^-\Sigma^+$
(c) $\pi^-p \to \overline{\Sigma}^-\Sigma^0 p$
(d) $\overline{p}p \to \pi^+\pi^+\pi^-\pi^-$
(e) $\pi^+p \to K^+\Sigma^+$
[See solutions]

Solutions

Problem 6.1 Consider a particle with spin (see Fig. 6.1); the spin defines a direction in space. The particle can be neutral but with a different distribution of positive and negative electric charges, so that the particle has a classical electric dipole moment $\mathbf{d}_E$. Since the particle is rotating, it also has a magnetic dipole moment $\mathbf{d}_M$. The same argument is valid for negatively or positively charged particles, with an asymmetric charge distribution.

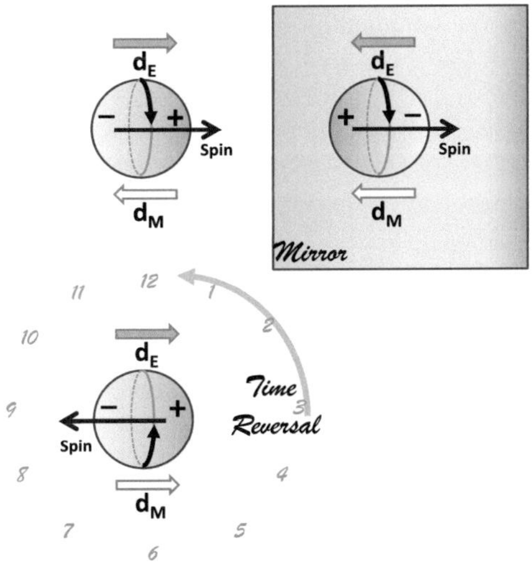

Fig. 6.1 A neutral fermion (the neutron, for instance) is represented here as a spherical object with an asymmetric charge density (*upper left*). The particle has a magnetic dipole moment $\mathbf{d}_M$ (here antiparallel to the spin). The particle mirror image is represented on the *right*, and its time-reversal image at the *bottom*. The particle spin defines a direction in space. Both P and T transformations leave the magnetic dipole moment antiparallel to the spin. Both P and T transformations change the relative orientation of the electric dipole moment. Therefore, the original particle can be distinguished from its mirror or its time reversal image

The original particle (left upper part of Fig. 6.1) has spin parallel to the electric dipole moment $\mathbf{d}_E$, and spin antiparallel (i.e., parallel but with opposite directions) to the magnetic dipole moment $\mathbf{d}_M$. By performing a parity inversion (right upper part of Fig. 6.1), the electric dipole moment $\mathbf{d}_E$ and spin are now antiparallel, whereas the magnetic dipole moment and spin remain antiparallel. Thanks to the electric dipole moment $\mathbf{d}_E$, it is therefore possible to distinguish the original particle from its mirror image, and the parity conservation is violated.

If we now perform a time reversal transformation, as shown in the bottom part of Fig. 6.1, the particle will spin in the opposite direction with respect to the original particle and the magnetic dipole moment $\mathbf{d}_M$ changes its direction as well. For the neutron, both the magnetic dipole moment $\mathbf{d}_M$ and the spin invert their directions and therefore, they remain antiparallel. The electric dipole moment $\mathbf{d}_E$ remains unchanged but since the spin has changed direction, they are now antiparallel to each other. There is a clear difference and the original and time reversal pictures can be distinguished.

In conclusion, if the particle has a non vanishing electric dipole moment, both the time reversal and parity invariance are violated [P87].

Problem 6.2 The application of the operators P and C to p and e^+ gives no particular problem. The problem arises instead for the $\overline{\nu}_e$. The parity operator transforms a right-handed $\overline{\nu}_e$ into a left-handed $\overline{\nu}_e$, which does not exist. After the application of P, the charge conjugation operator transforms the left-handed $\overline{\nu}_e$ into a left-handed ν_e, which has the correct helicity. Only the consecutive applications of the two operators P and C is therefore possible. The application of CP to the neutron β decay produces the possible process $\overline{n} \to \overline{p}e^- \nu_e$.

Problem 6.3

(a) $\tau = \hbar/\Gamma = \frac{6.6 \cdot 10^{-22}\ \mathrm{MeV\,s}}{1.3 \cdot 10^{-3}\ \mathrm{MeV}} = 5.1 \cdot 10^{-19}$ s. This is a lifetime typical of the electromagnetic interaction, as shown also by the fact that the decay into two photons has the largest BR.
(b) Forbidden decays.

- $\eta^0 \to \gamma\gamma\gamma$. The decay into 3 γ's has a charge conjugation eigenvalue $C = (-1)^3$. The η^0 has $C = +1$. The decay into 2 photons is allowed, while the decay into 3 photons is forbidden because the electromagnetic interaction conserves C.
- $\eta^0 \to \pi^0\pi^0$. The decay into 2 π^0's has a parity eigenvalue $P = (-1)^2$. The η^0 has $P = -1$. The decay into 3 pions is allowed, while the decay into 2 pions is forbidden because the electromagnetic interaction conserves P.
- $\eta^0 \to \pi^0\gamma$. The η^0 has spin zero, while the final state has spin 1. The decay is forbidden by angular momentum conservation (verify that also C is violated).
- $\eta^0 \to e^+\mu^-$ or $e^-\mu^+$. These decays are forbidden by lepton flavor number conservation.

Problem 6.4 In terms of the valence quarks, the two Σ^0 and Λ^0 baryons have the same (uds) composition. The Σ^0 represents the neutral particle of a $I = 1$ strong isospin triplet state, while the Λ^0 is an isospin singlet, with $I = 0$. The Σ^0 decay into a Λ^0 is forbidden in the strong interaction (which conserves the isospin), but it is possible in the electromagnetic interaction (see Table 6.4). The decay into $n\pi^0$ is a strangeness violating decay, which is forbidden both in strong and electromagnetic interactions.

Problem 6.5 The Ω^- particle is a three s quark state and has strangeness $S = -3$. Both the strong and electromagnetic interactions conserve the quark flavor and therefore, the strangeness. The only allowed decays are through the weak interaction. Note that the lifetimes of the decay products are long enough so that their tracks are visible in bubble chamber pictures (see Problem 8.11)

Problem 6.6

(a) Consider the decay of N_0 neutral pions. If the π^0 has spin 0, the angular distribution of the emitted photons in the c.m. system must be isotropic, since there is no preferential direction:

$$\frac{dN}{d\Omega^*} = \frac{N_0}{4\pi} \tag{6.1}$$

where $d\Omega^*$ is the infinitesimal solid angle in the c.m. system. From (6.1), one derives:

$$dN = \frac{N_0}{4\pi} d\Omega^* = \frac{N_0}{4\pi} 2\pi \, d\cos\theta^* \longrightarrow \frac{dN}{d\cos\theta^*} = \frac{N_0}{2}. \tag{6.2}$$

The angle θ^* in the c.m. system is that formed between the direction of the considered photon with the direction of the pion in the laboratory system. If the pion had non-null spin, its projection along the direction of motion would give rise to structures in the angular distribution of the emitted photons.

(b) In the pion rest system, independently of the emission angle, each photon has an energy equal to $E^*_\gamma = \frac{m_\pi}{2}$ and a momentum $\mathbf{p}^*_\gamma$ opposite each other. Let us now express the energy of the photon in the laboratory system. In the lab. system, the pion energy-momentum four-vector is equal to $(E_\pi, \mathbf{p}_\pi)$, and the two reference systems have a relative Lorentz boost equal to:

$$\beta = \frac{|\mathbf{p}_\pi|}{E_\pi}; \qquad \Gamma = \frac{E_\pi}{m_\pi}.$$

(Note that the capital letter Γ is used for the Lorentz factor $\Gamma = 1/\sqrt{1-\beta^2}$, to avoid confusion with the symbol γ of emitted photons.) As β, Γ are known, the photon energy can be determined in the lab. system (remembering that this system is seen by the pion with a negative velocity) using the Lorentz transformation:

$$E_\gamma = \Gamma E^*_\gamma + \beta \Gamma p^*_\gamma \cdot \cos\theta^* \tag{6.3}$$

from which one derives:

$$E_\gamma = \frac{E_\pi}{m_\pi} \frac{m_\pi}{2} \left(1 + \beta\cos\theta^*\right) \longrightarrow E_\gamma = \frac{E_\pi}{2} \left(1 + \beta\cos\theta^*\right) \tag{6.4}$$

having used the relation $p^*_\gamma = E^*_\gamma$.

Note that depending on the angle of emission in the c.m., the photon energy in the laboratory can vary between:

$$E_\gamma^{min} = \frac{E_\pi}{2}(1-\beta); \qquad E_\gamma^{max} = \frac{E_\pi}{2}(1+\beta). \tag{6.5}$$

The shape of the energy spectrum of emitted photons in the laboratory frame, dN/dE_γ, is obtained by differentiating Eq. (6.3) with respect to the variable $\cos\theta^*$:

$$dE_\gamma = \beta \Gamma p^*_\gamma \cdot d\cos\theta^* \longrightarrow \frac{d\cos\theta^*}{dE_\gamma} = \frac{1}{\beta \Gamma p^*_\gamma} \tag{6.6}$$

from which one finds:

$$\frac{dN}{dE_\gamma} = \frac{dN}{d\cos\theta^*} \frac{d\cos\theta^*}{dE_\gamma} = \frac{N_0}{2} \frac{1}{\beta \Gamma p^*_\gamma}. \tag{6.7}$$

All the quantities in the right member of Eq. (6.6) are constant, which means that the distribution of the number of photons is constant in the range $[E_\gamma^{min}, E_\gamma^{max}]$.

(c) For a π^0 with energy $E_\pi = 0.8$ GeV, one has:

$$\beta = \frac{p_\pi}{E_\pi} = \frac{\sqrt{E_\pi^2 - m_\pi^2}}{E_\pi} = 0.9856$$

From Eq. (6.5), one finds $E_\gamma^{min} = 7.2 \cdot 10^{-3} E_\pi$; $E_\gamma^{max} = 0.9920 E_\pi$.

Problem 6.7 Reaction (b) is forbidden as the final state has strangeness quantum number $S = -2$, while the initial state is not strange.
Reaction (d) cannot occur through the strong interaction. The antiparticles have the same isospin of particles: hence the initial state ($\overline{p}p$) has isospin $I = 1$. In the final state, there are 4 particles with isospin $I = 1$.

Chapter 7
Interactions of Hadrons at Low Energies and the Static Quark Model

Problems

7.1. **Range of the nuclear force**. Despite the fact that the quarks within a neutron or a proton interact via the exchange of gluons, the strong interaction between a proton and a neutron can be viewed as due to the exchange of a pion. If the π has a mass of 140 GeV/c^2, calculate the distance at which the strong force is effective.
[See solutions]

7.2. **Λ^0 decay at rest**. For the $\Lambda^0 \rightarrow p\pi^-$ decay at rest, calculate the momentum and the energy of the final state particles.
[See solutions]

7.3. **π^- interactions**. A target of liquid hydrogen (density $\rho = 0.06$ g cm^{-3}) has a volume of 100 cm^3. A monoenergetic π^- beam with 300 MeV/c momentum collides on the target. The beam is broad, uniform and with an intensity of $\Phi = 10^7\ \pi^-\ \mathrm{m}^{-2}\ \mathrm{s}^{-1}$. The cross-section for the reaction $\pi^- p \rightarrow \pi^0 n$ at 300 MeV/c is $\sigma = 45$ mb.
(a) Calculate the number of γ-rays produced per second (recall that the π^0 meson decays into two photons, $\pi^0 \rightarrow 2\gamma$, in a very short time).
(b) Calculate the mean free path of 300 MeV/c π^- in liquid hydrogen.
[See solutions]

7.4. **Antiproton capture at rest**. The antiprotons captured at rest in deuterium give rise to the reaction $\overline{p}d \rightarrow n\pi^0$. Determine:
(a) the deuterium binding energy;
(b) the total energy of the emitted π^0 meson;
(c) the process that occurs in terms of valence quarks.
[See solutions]

S. Braibant et al., *Particles and Fundamental Interactions: Supplements, Problems and Solutions*, Undergraduate Lecture Notes in Physics,
DOI 10.1007/978-94-007-4135-5_7, © Springer Science+Business Media Dordrecht 2012

7.5. **Energy thresholds**. Consider the collisions of $\overline{K}^0$ and K^0 mesons on protons at rest.

(a) Calculate, in the laboratory system, the minimum kinetic energy of the $\overline{K}^0$ and of the K^0, necessary to induce respectively the following reactions:

$$\overline{K}^0 + p \to \Lambda^0 + \pi^+ \tag{7.1}$$

$$K^0 + p \to \Lambda^0 + K^0 + K^+ \tag{7.2}$$

(b) With the help of Table 7.3, write the reactions in terms of valence quarks.

[See solutions]

7.6. **ϕ(1020) decay**. Explain why the $\phi(1020)$ vector meson cannot decay into two π^0 mesons.

[See solutions]

7.7. **ψ decay**. Explain why the $\psi(3685) \to J/\psi(3097)\pi^0$ decay is not allowed by the strong interaction, while the $\psi(3685) \to J/\psi(3097)\eta$ decay is permitted.

[See solutions]

7.8. **Δ^0 decay**. The $\Delta^0[udd]$ resonance decays mainly in $p\,\pi^-$ with a width $\Gamma \simeq$ 100 MeV. Draw the Feynman diagram. Estimate the Δ^0 lifetime. Is there another possible decay channel?

[See solutions]

7.9. **The Δ^{++} resonance**. Calculate the elastic $\pi^+ p$ cross-section for the formation of the Δ^{++} (1232) resonance when

(a) the π^+ in the lab. system has a kinetic energy $T_{\pi_{lab}} = 190$ MeV;

(b) the π^+ in the lab. system has a kinetic energy $T_{\pi_{lab}} = 300$ MeV.

Assume natural units and ($m_p = 938, m_\pi = 140, m_\Delta = 1232, \Gamma = 120$) MeV.

[See solutions]

7.10. **Mean free path**. For a graphite target ($\rho = 2.265$ g cm^{-3}), calculate the number N_n of carbon nuclei per cm^3, the absorption coefficient μ for a proton beam and the interaction length λ. Use a cross-section value of $\sigma = 0.331 \cdot 10^{-24}$ cm^2.

[See solutions]

7.11. **Isospin-1: pion-proton collisions**. Express, as a function of the isospin, the amplitudes of the following collision processes:

(a) $\pi^+ p \to \pi^+ p$

(b) $\pi^- p \to \pi^- p$

(c) $\pi^- p \to \pi^0 n$
(d) Calculate the ratio between the cross-sections of the above three processes, assuming that the $I = 3/2$ isospin channel is dominant. Explain the physical motivation for this assumption.

Refer to Supplement 7.1 and Figs. 7.14, 7.15, 7.17 and 7.18 for the isospin values of the involved particles.
[See solutions]

7.12. **Isospin-2: Δ resonance formation in pion-proton collisions**. The Δ resonance is an $I = 3/2$ isospin multiplet. Consider the following two formation mechanisms

$$\pi^- p \to \Delta^0$$
$$\pi^+ p \to \Delta^{++}$$

and determine the ratio between the respective cross-sections.
[See solutions]

7.13. **Isospin-3: $K^- p$ collisions**. Calculate the cross-section ratio of the following processes:
(a) $K^- p \to \pi^+ \Sigma^-$
(b) $K^- p \to \pi^0 \Sigma^0$
(c) $K^- p \to \pi^- \Sigma^+$
(d) $K^- p \to \pi^0 \Lambda^0$
[See solutions]

7.14. **Isospin-4: the *pd* cross-section**. Calculate the cross-section ratio for the reactions $pd \to {}^3\text{He}\, \pi^0$, $pd \to {}^3\text{H}\, \pi^+$ at a fixed energy in the c.m. system.
[A. $\sigma(pd \to {}^3\text{He}\, \pi^0)/\sigma(pd \to {}^3\text{H}\, \pi^+) \simeq 1/2$]

7.15. **Isospin-5: Strange and Charmed particles**. Indicate which is the isotopic spin of the π^-, K^-, D^0, D_s^+ mesons and of Σ^-, Σ_c^0, Ξ_{cc}^+ and Ω_{cc}^+ baryons.
[Hint: Refer to Figs. 7.17, 7.18]

7.16. **Weisskopf formula for neutral vector meson decay**. Starting from Eq. (4.17):

$$\Gamma = \frac{16\pi\alpha^2}{m_e^2}|\psi(0)|^2 \tag{7.3}$$

derive the Weisskopf formula Eq. (7.59) for the decay of the ρ^0, ω^0, ϕ^0 neutral vector mesons into leptons.
[See solutions]

7.17. **ρ^0 Spin-parity**. Determine the spin-parity of the ρ^0 resonance produced in $\pi^- p \to \rho^0 n$, with the subsequent decay into two pions, $\rho^0 \to \pi^- \pi^+$.

In the above reaction a peak is observed at the invariant mass $m_\rho = m_{\pi^+\pi^-} = 775$ MeV, with a $\Gamma = 149$ MeV width. Explain why no $\pi^0\pi^0$ resonance is observed at the same mass.

[See solutions]

Supplement 7.1: Sum of Angular Momentum and Isospin: the Clebsch–Gordan Coefficients

The formalism of angular momentum composition is widely used in quantum-mechanics and is described in any textbook. In particle physics, the same formalism is used in processes involving interactions of hadrons, made of u, d quarks, through the *isotopic spin* (or *isospin*) (see Sect. 7.2) composition. The strong interaction depends on the isospin I, but not on the component along the quantization axis (conventionally, the z-axis) I_z. Proton and neutron belong to an $I = 1/2$ isospin doublet with $I_z = +1/2, -1/2$, respectively. For the strong interaction, the proton and the neutron are two degenerate states. The isotopic spin of mesons and hadrons made of u, d quarks is presented in Figs. 7.14, 7.15, 7.17 and 7.18.

Suppose to have two particles with isospin $\mathbf{I}_1$ and $\mathbf{I}_2$ with respective components along the quantization axis, I_{z_1} and I_{z_2}. The isospin composition helps, for instance, to evaluate the probability that, in $\pi^- p$ collisions, a $\pi^- p$ or a $\pi^0 n$ state be formed (Problem 7.11). We need to calculate the eigenvalue of the sum $\mathbf{I} = \mathbf{I}_1 + \mathbf{I}_2$ in a way that the projection along the z-axis be the sum of the particle third components: $I_z = I_{z_1} + I_{z_1}$.

Let us define:

- $|I_1, I_{z_1}; I_2, I_{z_2}\rangle$: the eigenstate which describes the sum of the two particles, with the properties:

$$I_1^2 |I_1, I_{z_1}; I_2, I_{z_2}\rangle = I_1(I_1+1)|I_1, I_{z_1}; I_2, I_{z_2}\rangle \tag{7.4a}$$

$$I_{z_1} |I_1, I_{z_1}; I_2, I_{z_2}\rangle = m_1 |I_1, I_{z_1}; I_2, I_{z_2}\rangle \tag{7.4b}$$

$$I_2^2 |I_1, I_{z_1}; I_2, I_{z_2}\rangle = I_2(I_2+1)|I_1, I_{z_1}; I_2, I_{z_2}\rangle \tag{7.4c}$$

$$I_{z_2} |I_1, I_{z_1}; I_2, I_{z_2}\rangle = m_2 |I_1, I_{z_1}; I_2, I_{z_2}\rangle \tag{7.4d}$$

- $|I, M\rangle$: the eigenstate corresponding to the sum in the new basis in which I^2 and I_z are diagonal:

$$I^2|I, M\rangle = I(I+1)|I, M\rangle; \qquad I_z|I, M\rangle = M|I, M\rangle \tag{7.5}$$

The eigenstates $|I, M\rangle$ can be expressed as a linear combination of the initial eigenstates:

$$|I, M\rangle = \sum_{m_1} \sum_{m_2} |I_1, I_{z_1}; I_2, I_{z_2}\rangle \langle I_1, I_{z_1}; I_2, I_{z_2} | I, M\rangle \tag{7.6}$$

The scalar products $\langle I_1, I_{z_1}; I_2, I_{z_2} | I, M\rangle$ are the so-called *Clebsch-Gordan coefficients*, and are reported in Fig. 7.1 for different combinations of I_1, I_2. The eigenvalues of the sum must fulfill the conditions:

$$|I_1 - I_2| \le I \le |I_1 + I_2|; \qquad M = m_1 + m_2 = I_{z_1} + I_{z_2} \tag{7.7}$$

Problems from 7.11 to 7.15 help to understand the use of the Clebsch-Gordan coefficients in different situations.

Solutions

Problem 7.1 The uncertainty principle states that:

$$\Delta E \cdot \Delta t \ge \hbar \longrightarrow \Delta t \ge \frac{\hbar}{\Delta E}.$$

The distance traveled in a time Δt is:

$$\Delta r = c \cdot \Delta t.$$

Therefore, the interaction effective range is given by:

$$\Delta r \simeq \frac{\hbar c}{\Delta E} = \frac{\hbar c}{mc^2} = \frac{197 \text{ MeV fm}}{140 \text{ MeV}} = 1.4 \cdot 10^{-15} \text{ m} = 1.4 \text{ fm}.$$

Problem 7.2 The system in which the Λ baryon is at rest coincides with the center of mass system. The momenta of the two particles in the final state must be equal in magnitude: $p_\pi = p_p$. Therefore, one has:

$$s = (E_\pi + E_p, \mathbf{p}_\pi + \mathbf{p}_p)^2 = m_\pi^2 + m_p^2 + 2E_\pi E_p + 2p_\pi^2 \tag{7.8}$$

from which, one finds:

$$p_\pi^2 = \frac{s - m_\pi^2 - m_p^2 - 2E_\pi E_p}{2} \tag{7.9}$$

The relativistic invariant $\sqrt{s}$ is the total energy of the system, and $E_p = \sqrt{s} - E_\pi$. Substituting E_p in Eq. (7.9), one can write:

$$p_\pi^2 = \frac{s - m_\pi^2 - m_p^2 - 2E_\pi(\sqrt{s} - E_\pi)}{2}$$

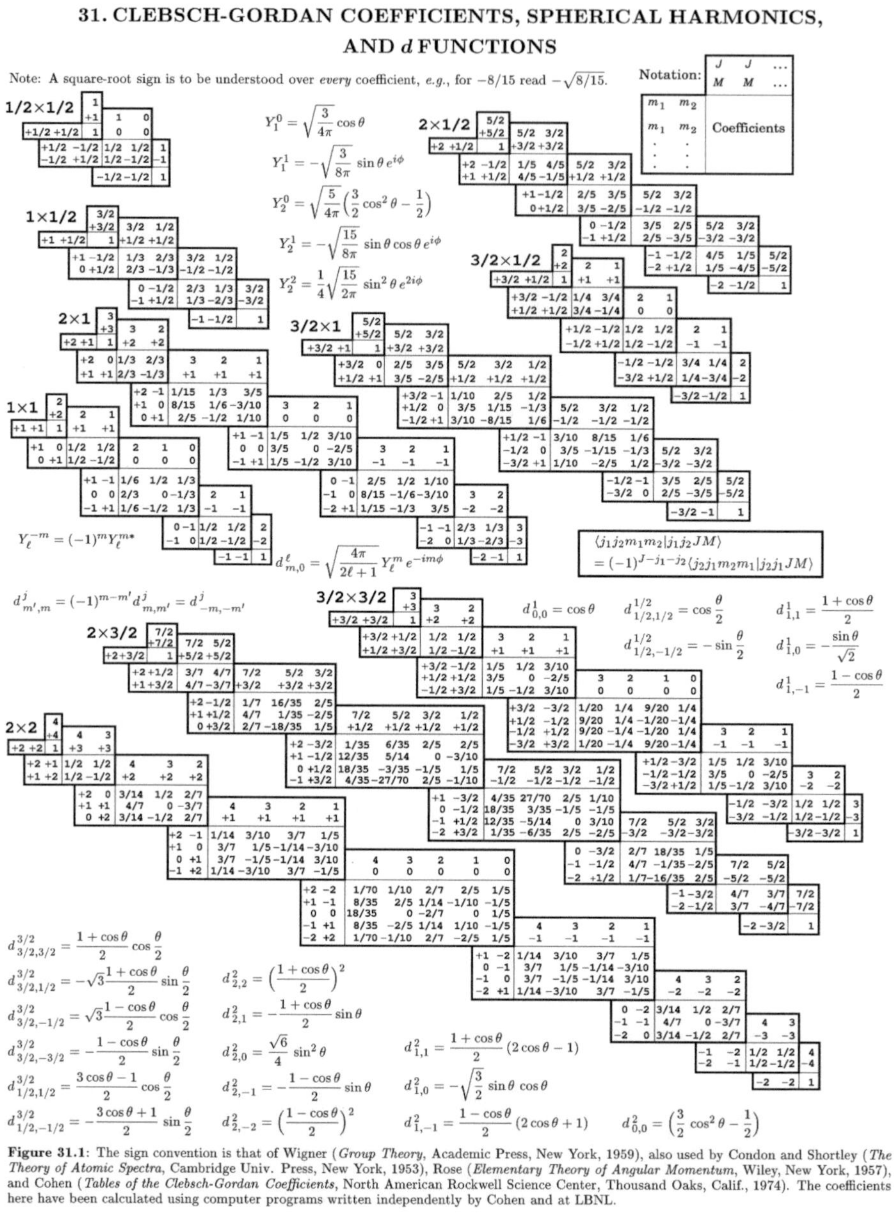

Fig. 7.1 Clebsch-Gordan coefficients from [P10]

from which, one has:

$$E_\pi = \frac{s - m_p^2 + m_\pi^2}{2\sqrt{s}}; \qquad E_p = \frac{s - m_\pi^2 + m_p^2}{2\sqrt{s}} \tag{7.10}$$

By inserting the numerical values ($c=1$): $\sqrt{s}=m_\Lambda=1115$ MeV; $m_\pi=140$ MeV; $m_p=938$ MeV, one obtains:

$$E_\pi = 170 \text{ MeV}; \qquad E_p = 943 \text{ MeV}$$

$$p_\pi = \sqrt{E_\pi^2 - m_\pi^2} = 97 \text{ MeV/c}; \qquad p_p = \sqrt{E_p^2 - m_p^2} = 97 \text{ MeV/c}$$

Problem 7.3 The number of interactions per second (i.e., the number of π^0 produced per second) is given by $R_{\pi^0} = \Phi\sigma N$, where N is the number of scattering centers in the given volume $V = 100\ \text{cm}^3$. Then: $N = \rho N_A V/A = 3.6 \cdot 10^{24}\ \text{cm}^3$, where N_A is Avogadro's number and $A = 1$ for hydrogen. Since the π^0 decays into two photons, the number of γ's produced per second is:

$$N_\gamma = 2R_{\pi^0} = 2\,\Phi\sigma N = 2\times 10^3\ \text{cm}^{-2}\,\text{s}^{-1} \times 45\cdot 10^{-27}\ \text{cm}^2 \times 3.6\cdot 10^{24} = 324\ \text{s}^{-1}.$$

The mean free path is $\lambda = A/(N_A\sigma) = 37\ \text{g}\,\text{cm}^{-2}$.

Problem 7.4

(a) The masses of the involved particles are: $m_{\overline{p}} = m_p = 938.3$ MeV, $m_n = 939.6$ MeV, $m_d = 1875.6$ MeV, $m_{\pi^0} = 140$ MeV. The binding energy is: $\text{BE} = m_d - (m_p + m_n) = 2.3$ MeV.

(b) The π^0 energy can be determined from Eq. (3.11) with $\sqrt{s} = m_d + m_{\overline{p}}$:

$$E_\pi^* = \frac{s - m_n^2 + m_\pi^2}{2\sqrt{s}}. \tag{7.11}$$

Inserting numerical values, one finds $E_\pi = 1253.5$ MeV.

(c) In terms of constituent quarks, the antiproton composition is: $\overline{p} = [\overline{u}\overline{u}\overline{d}]$; the deuterium nucleus is made of a proton and a neutron: $(pn) = [(uud), (udd)]$. The interaction process, due to the strong interaction, consists of the annihilation of a $\overline{u}\overline{d}$ pair of the antiproton with a ud pair of the proton/neutron. The remaining $(\overline{u}u)$ pair forms the outgoing π^0.

Problem 7.5 The masses of particles involved in the reaction are expressed in MeV. For the reaction (7.1), one has:

$$\begin{array}{ccccccc} \overline{K}^0 & + & p & \to & \Lambda & + & \pi^+ \\ 498 & & 938 & & 1116 & & 140 \text{ MeV}. \end{array}$$

The sum of the masses of the initial state is 1436 MeV; the sum of final masses is 1256 MeV, smaller than the initial one. The reaction can always take place and kinetic energy of initial particles is not required (the reaction is above the threshold). For reaction (7.2), one has:

$$\begin{array}{ccccccccc} K^0 & + & p & \to & \Lambda & + & K^0 & + & K^+ \\ 498 & & 938 & & 1116 & & 498 & & 494 \text{ MeV}. \end{array}$$

The sum of the final state masses is $m_f = m_\Lambda + m_{K^0} + m_{K^+} = 2108$ MeV. Assuming that the proton is at rest, the minimum kinetic energy $T = E_{K^0} - m_{K^0}$ required for the incoming K^0 can be determined using the energy-momentum relativistic invariant s. The energy-momentum of the initial state is:

$$\left[(E_{K^0}, \mathbf{p}_{K^0}) + (m_p, 0)\right] \tag{7.12}$$

and its relativistic invariant s is:

$$s = m_{K^0}^2 + m_p^2 + 2m_p E_{K^0} \tag{7.13}$$

For the final state, assuming that the particles are produced at rest:

$$s = (m_\Lambda + m_{K^0} + m_{K^+})^2 = m_f^2 \tag{7.14}$$

By equating Eq. (7.13) with Eq. (7.14), one obtains:

$$E_{K^0} = \frac{m_f^2 - m_p^2 - m_{K^0}^2}{2m_p} = \frac{2108^2 - 938^2 - 498^2}{2 \cdot 938} = 1770 \text{ MeV}. \tag{7.15}$$

The reaction can take place only if the kinetic energy of the incoming K^0 is larger than $T = E_{K^0} - m_{K^0} = 1770 - 498 \simeq 1270$ MeV.

(b) In the second reaction, a $(s\bar{s})$ pair must be created by the strong interaction:

$$\overline{K}^0 p = [s\bar{d}][uud] \to [uds][\bar{d}u] = \Lambda^0 \pi^+ \tag{7.16}$$

$$K^0 p = [d\bar{s}][uud] \to [d\bar{s}](s\bar{s})[uud] = [d\bar{s}][sud][u\bar{s}] = K^0 \Lambda^0 K^+ \tag{7.17}$$

Problem 7.6 The ϕ meson has spin 1, the π^0 meson has spin 0. The conservation of momentum in the hypothetical $\phi \to \pi^0\pi^0$ decay would imply a relative orbital moment $\ell = 1$ (P-wave). Recall that the bosons follow the *Bose–Einstein statistics* and that the wave function of a system of identical bosons must be symmetric for the exchange of any two bosons. The P-wave decay is therefore forbidden by the Bose-Einstein statistics since a system of two identical bosons $(\pi^0\pi^0)$ in the P-wave implies an antisymmetric wave function.

With the same reasoning, the possibility for the $Z^0 \to H^0 H^0$ decay is excluded, assuming that the H^0 Higgs boson has spin zero. All decays of a vector boson $(J = 1)$ in two identical scalar particles $(J = 0)$ are excluded, for example $\rho^0 \to \pi^0\pi^0$.

Problem 7.7 The initial state $\psi(3685)$ consists of a pair of $c\bar{c}$ quarks: it is an isospin singlet and its isospin (quantum number different from zero only for u, d quarks) is $I = 0$. The final state J/ψ has the same isospin assignment.

The π^0 is the neutral particle of a $I = 1$ isospin triplet; its third component is $I_z = 0$. The η is an isospin singlet with $I = I_z = 0$.

In the following decay:

$$\begin{array}{lcc} & \psi(3685) \rightarrow & J/\psi \ \pi^0 \\ \text{isospin } 0 & & 0 \quad 1, \end{array} \tag{7.18}$$

the isospin is not conserved and the decay is forbidden for the strong interaction. In the second decay:

$$\begin{array}{lcc} & \psi(3685) \rightarrow & J/\psi \ \eta^0 \\ \text{isospin } 0 & & 0 \quad 0, \end{array} \tag{7.19}$$

the isospin is conserved and the decay is allowed for the strong interaction.

Problem 7.8 The Δ^0 [udd] resonance decays mainly in $p\,\pi^-$.

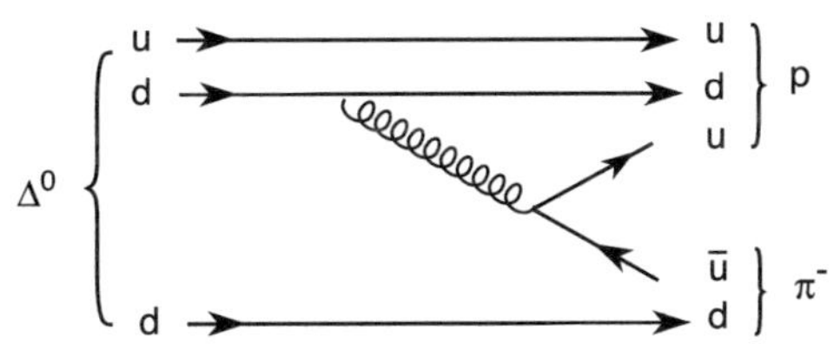

The width of the Δ^0 resonance is $\Gamma \simeq 100$ MeV. With $\Delta E \simeq \Gamma$ and $\Delta t \simeq \tau$, and with the uncertainty principle $\Delta E \cdot \Delta t \geq \hbar$, one gets:

$$\tau \geq \hbar/\Gamma = \frac{6.6 \cdot 10^{-22}\ \text{MeV s}}{100\ \text{MeV}} = 6.6 \cdot 10^{-24}\ \text{s} \simeq 10^{-23}\ \text{s}.$$

Another possible strong interaction decay is $\Delta^0 \rightarrow n\pi^0$.

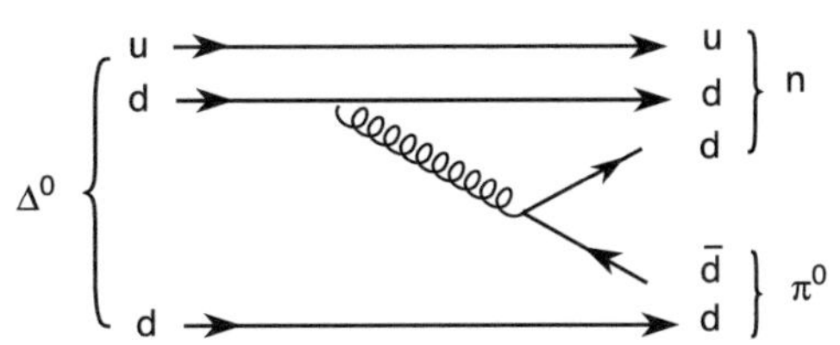

Problem 7.9 The Breit-Wigner resonance formula for the Δ^{++} with spin 3/2 is given in Eq. (7.27). The quantities are expressed in the c.m. system and the energy and momentum must therefore be calculated in this system. In a first step, the energy-momentum four-vector in the laboratory system is determined knowing $T_{\pi_{lab}}$; in a second step, the c.m. system values are obtained.

In the laboratory system, the pion energy-momentum four-vector is $(E_{\pi lab}, \mathbf{p}_{\pi lab})$, with $E_{\pi lab} = m_\pi + T_{\pi lab}$, and that of the proton at rest is $(m_p, 0)$. The relativistic invariant is:

$$s = [(E_{\pi lab}, \mathbf{p}_{\pi lab}) + (m_p, 0)]^2 = E_{\pi lab}^2 + m_p^2 + 2E_{\pi lab} m_p - p_{\pi lab}^2$$
$$= m_\pi^2 + m_p^2 + 2E_{\pi lab} m_p$$

from which, one finds:

$$E_{\pi lab} = \frac{s - m_\pi^2 - m_p^2}{2m_p} \tag{7.20}$$

The pion energy in the laboratory system corresponding to $\sqrt{s} = m_\Delta = 1232$ MeV is obtained from Eq. (7.20):

$$E_{\pi lab} = 330 \text{ MeV}; \qquad T_{\pi lab} = E_{\pi lab} - m_\pi = 190 \text{ MeV}$$

which corresponds precisely to the first kinetic energy value given in the problem.

The pion momentum p_π^* can now be evaluated in the c.m. system, where the π and proton energy-momentum four-vectors are respectively, $(E_\pi^*, \mathbf{p}_\pi^*)$ and $(E_p^*, \mathbf{p}_p^*)$, with the condition $\mathbf{p}_\pi^* + \mathbf{p}_p^* = 0$. The relativistic invariant in the c.m. system is:

$$s = \left(E_\pi^* + E_p^*\right)^2 = m_\pi^2 + p_\pi^{*\,2} + m_p^2 + p_p^{*\,2} + 2E_\pi^* E_p^*$$
$$= \left[m_\pi^2 + m_p^2 + 2p_\pi^{*\,2} + 2E_\pi^*\left(\sqrt{s} - E_\pi^*\right)\right]$$
$$= \left[m_\pi^2 + m_p^2 + 2\left(p_\pi^{*\,2} - E_\pi^{*\,2}\right) + 2E_\pi^*\sqrt{s}\right]$$
$$= \left[m_p^2 - m_\pi^2 + 2E_\pi^*\sqrt{s}\right]$$

from which, one gets:

$$E_\pi^* = \frac{s - m_p^2 - m_\pi^2}{2\sqrt{s}}; \qquad p_\pi^* = \sqrt{E_\pi^{*\,2} - m_\pi^2} \tag{7.21}$$

(a) For $T_{\pi lab} = 190$ MeV $\rightarrow E_{\pi lab} = 330$ MeV $\rightarrow \sqrt{s} = m_\Delta = 1232$ MeV and from Eq. (7.21), one finds:

$$E_\pi^* = \frac{m_\Delta^2 - m_p^2 - m_\pi^2}{2\sqrt{m_\Delta}} = 267 \text{ MeV}; \qquad p_\pi^* = \sqrt{E_\pi^{*\,2} - m_\pi^2} = 228 \text{ MeV}$$

The cross-section has a maximum corresponding to:

$$\sigma(T_{\pi lab} = 190 \text{ MeV}) = \sigma_{max} = \frac{8\pi(\hbar c)^2}{p_\pi^{*\,2}} = \frac{8\pi(0.389 \text{ GeV}^2 \text{ mb})}{(0.228 \text{ GeV})^2} = 188 \text{ mb} \tag{7.22}$$

(b) For $T_{\pi_{lab}} = 300$ MeV $\to E_{\pi_{lab}} = 440$ MeV $\to \sqrt{s} = \sqrt{m_\pi^2 + m_p^2 + 2E_{\pi_{lab}} m_p}$ $= 1313$ MeV and from Eq. (7.21), one finds:

$$E_\pi^* = \frac{s - m_p^2 - m_\pi^2}{2\sqrt{s}} = 329 \text{ MeV}; \qquad p_\pi^* = \sqrt{E_\pi^{*2} - m_\pi^2} = 298 \text{ MeV}$$

The corresponding value of the cross-section is:

$$\begin{aligned}\sigma(T_{\pi_{lab}} = 300 \text{ MeV}) &= \frac{8\pi(\hbar c)^2}{p_\pi^{*2}} \frac{\Gamma^2/4}{(m_\Delta - \sqrt{s})^2 + \Gamma^2/4} \\ &= \sigma_{max} \left(\frac{0.228 \text{ GeV}}{0.298 \text{ GeV}}\right)^2 \left(\frac{0.120^2/4}{(1.232 - 1.313)^2 + 0.120^2/4}\right) \\ &= \sigma_{max}(0.58)(0.35) = 0.2\sigma_{max}\end{aligned}$$

Problem 7.10 The number of scattering centers is $N_n = N_a = \frac{\rho}{A} N_A$:

$$\begin{aligned}&N_n = 1.137 \cdot 10^{23} \text{ (C nuclei per cm}^3) \\ &\mu = N_n \sigma = 1.137 \cdot 10^{23} \cdot 0.331 \cdot 10^{-24} \\ &\quad = 3.76 \cdot 10^{-2} \text{ (nuclei per cm of thickness)} \\ &\lambda \text{ (cm)} = 1/\mu = 26.5 \text{ cm}, \qquad \lambda\,(\text{g cm}^{-2}) = \rho/\mu = 60.2 \text{ g cm}^{-2}.\end{aligned}$$

Note that μ also represents the surface fraction covered with carbon nuclei ($\simeq$3.76%) for a target 1 cm thick.

From Eq. (7.8), one obtains a simple practical formula:

$$\left| -\frac{dI}{I} \right| = \sigma \cdot \frac{\rho}{A} N_A n \cdot dx \simeq \sigma\, 0.6\, \frac{n}{A}$$

where σ is expressed in barn, the thickness ρdx in g cm^{-2}, A is the atomic weight and n is the number of targets per atom (in this case $n = 1$; if the collision arises on electrons for example, one has $n = Z$).

Problem 7.11 All mentioned processes are due to the strong interaction which conserves the isospin. The amplitudes of initial and final states can be parameterized in terms of isospin values. Transitions occur only between states of the same isospin. The transition amplitudes are based on internal mechanisms of the strong interaction and depend on the isospin values.

(a) $\pi^+ p \to \pi^+ p$
In terms of isospin (states denoted as $|I, I_z\rangle$), one has: $|\pi^+\rangle = |1, 1\rangle$ and $|p\rangle = |\frac{1}{2}, \frac{1}{2}\rangle$. The combination of the third component produces $M = 3/2$ and the

only possible total isospin value is $I = 3/2$. The amplitude of the transition probabilities $\pi^+ p \to \pi^+ p$ depends on an isospin-dependent operator $A_{3/2}$:

$$\langle \pi^+ p | A_{3/2} | \pi^+ p \rangle = \left\langle \frac{3}{2}, \frac{3}{2} \right| A_{3/2} \left| \frac{3}{2}, \frac{3}{2} \right\rangle = a_{3/2} \tag{7.23}$$

(b) $\pi^- p \to \pi^- p$
In terms of isospin $|\pi^-\rangle = |1, -1\rangle$; the combination of a π^- with a proton $|p\rangle = |\frac{1}{2}, \frac{1}{2}\rangle$ has third component of the isospin along the quantization axis $M = -1 + 1/2 = -1/2$ corresponding to $1 - \frac{1}{2} \leq I \leq 1 + \frac{1}{2}$. The Clebsch-Gordan coefficients are obtained from the box $1 \times 1/2$ in Fig. 7.1, for the columns

$$\begin{matrix} J & \frac{3}{2} & \frac{1}{2} \\ M & -\frac{1}{2} & -\frac{1}{2} \end{matrix}$$

and for the line

$$m_1 \quad m_2 \quad -1 \quad \frac{1}{2}.$$

Recalling that a square root must be placed over every coefficient in the table, the Clebsch-Gordan coefficients are respectively for the first column: $\sqrt{\frac{1}{3}}$ and for the second column: $-\sqrt{\frac{2}{3}}$. The final state can therefore be written as the superposition:

$$|\pi^- p\rangle = \sqrt{\frac{1}{3}} \left| \frac{3}{2}, -\frac{1}{2} \right\rangle - \sqrt{\frac{2}{3}} \left| \frac{1}{2}, -\frac{1}{2} \right\rangle$$

The amplitude of the $\pi^- p \to \pi^- p$ transition probability is:

$$\begin{aligned} \langle \pi^- p | A | \pi^- p \rangle &= \frac{1}{3} \left\langle \frac{3}{2}, -\frac{1}{2} \right| A_{3/2} \left| \frac{3}{2}, -\frac{1}{2} \right\rangle + \frac{2}{3} \left\langle \frac{1}{2}, -\frac{1}{2} \right| A_{1/2} \left| \frac{1}{2}, -\frac{1}{2} \right\rangle \\ &= \frac{1}{3} a_{3/2} + \frac{2}{3} a_{1/2} \end{aligned} \tag{7.24}$$

(c) $\pi^- p \to \pi^0 n$
For the $\pi^- p$ interaction, the $\pi^0 n$ final state is also possible. These particles, in terms of isospin, are represented by: $|\pi^0\rangle = |1, 0\rangle$, $|n\rangle = |\frac{1}{2}, -\frac{1}{2}\rangle$. One has in this case $M = -1/2$ and $1/2 \leq I \leq 3/2$. The Clebsch-Gordan coefficients are obtained following the same procedure as that described above but for the line

$$m_1 \quad m_2 \quad 0 \quad -\frac{1}{2}.$$

The Clebsch-Gordan coefficients are respectively for the first column: $\sqrt{\frac{2}{3}}$ and for the second column: $\sqrt{\frac{1}{3}}$. The state is thus the superposition of:

$$|\pi^0 n\rangle = \sqrt{\frac{2}{3}}\left|\frac{3}{2}, -\frac{1}{2}\right\rangle + \sqrt{\frac{1}{3}}\left|\frac{1}{2}, -\frac{1}{2}\right\rangle$$

The amplitude of the $\pi^- p \to \pi^0 n$ transition probability is:

$$\langle\pi^0 n|A|\pi^- p\rangle = \sqrt{\frac{1}{3}}\sqrt{\frac{2}{3}}a_{2/3} - \sqrt{\frac{2}{3}}\sqrt{\frac{1}{3}}a_{1/2} = \sqrt{\frac{2}{9}}a_{3/2} - \sqrt{\frac{2}{9}}a_{1/2} \tag{7.25}$$

(d) The relation between the cross-sections depends on the square of the transition amplitude and on the phase-space factor. Since all the considered transitions are two-body decays into particles of (almost) the same mass, the phase-space factor is the same for the three considered processes. Therefore, one can write:

$$\sigma_a : \sigma_b : \sigma_c = a_{3/2}^2 : \frac{1}{9}|a_{3/2} + 2a_{1/2}|^2 : \frac{2}{9}|a_{3/2} - a_{1/2}|^2$$

One can assume that $a_{3/2} \gg a_{1/2}$ because the corresponding process with isospin $3/2$ can occur through the resonant reaction:

$$\pi^+ p \to \Delta^{++} \to \pi^+ p$$

which increases the cross-section. With this hypothesis, one obtains:

$$\sigma_a : \sigma_b : \sigma_c = a_{3/2}^2 : \frac{1}{9}a_{3/2}^2 : \frac{2}{9}a_{3/2}^2 = 1 : 1/9 : 2/9$$

If the $a_{1/2}$ term is not neglected, the values of $a_{3/2}, a_{1/2}$ can be estimated from the measured cross-section ratios. The calculation is an example of the predictive capabilities of the isospin formalism. With the only assumption that the interaction is governed by the strong interaction, a prediction concerning the cross-section values can be obtained.

Problem 7.12 This problem must be solved following the procedure and notation described in Problem 7.11. The strong interaction conserves the isospin of the combination of a proton $|p\rangle = |\frac{1}{2}, \frac{1}{2}\rangle$ plus a pion $|\pi^+\rangle = |1, 1\rangle$, $|\pi^-\rangle = |1, -1\rangle$.

The initial state is:

$$|\pi^- p\rangle = \sqrt{\frac{1}{3}}\left|\frac{3}{2}, -\frac{1}{2}\right\rangle - \sqrt{\frac{2}{3}}\left|\frac{1}{2}, -\frac{1}{2}\right\rangle.$$

The final state is represented by $|\Delta^0\rangle = |\frac{3}{2}, -\frac{1}{2}\rangle$.
The transition amplitude is therefore:

$$\langle\Delta^0|A|\pi^- p\rangle = \sqrt{\frac{1}{3}}\left\langle\frac{3}{2}, -\frac{1}{2}\right|A_{3/2}\left|\frac{3}{2}, -\frac{1}{2}\right\rangle = \sqrt{\frac{1}{3}}a_{3/2}.$$

In the second case, the initial state is $|\pi^+ p\rangle = |\frac{3}{2}, \frac{3}{2}\rangle$, the final one is $|\Delta^{++}\rangle = |\frac{3}{2}, \frac{3}{2}\rangle$ and the transition amplitude is:

$$\langle \Delta^{++}|A|\pi^+ p\rangle = \left\langle \frac{3}{2}, \frac{3}{2} \middle| A_{3/2} \middle| \frac{3}{2}, \frac{3}{2} \right\rangle = a_{3/2}$$

Since the phase-space factor is the same in both reactions, one can write:

$$\frac{\sigma(\pi^- p \to \Delta^0)}{\sigma(\pi^+ p \to \Delta^{++})} = \frac{|\langle \Delta^0|A|\pi^- p\rangle|^2}{|\langle \Delta^{++}|A|\pi^+ p\rangle|^2} = \frac{1}{3} \tag{7.26}$$

Let us compare the prediction with the experimental results, shown in Fig. 7.1. The resonant production of the Δ baryon occurs at $\sqrt{s} = 1230$ MeV. From the figure, one observes that the total cross-section of the $\pi^+ p$ process is $\sigma(\pi^+ p \to \Delta^{++}) \simeq 190$ mb (see also the analytic calculation given in Problem 7.9). The cross-section of the $\pi^- p$ process is $\sigma(\pi^- p \to \Delta^0) \simeq 65$ mb. Their ratio (65/190) is very close to the predicted ratio of Eq. (7.26).

Problem 7.13 The isospin (Figs. 7.14, 7.15) of the involved particles are (in addition to those already considered in the two previous problems): $|K^-\rangle = |\frac{1}{2}, -\frac{1}{2}\rangle$; $|\Sigma^+\rangle = |1, 1\rangle$; $|\Sigma^0\rangle = |1, 0\rangle$; $|\Sigma^-\rangle = |1, -1\rangle$; $|\Lambda^0\rangle = |0, 0\rangle$. The initial state is represented by:

$$|K^- p\rangle = \left|\frac{1}{2}, -\frac{1}{2}\right\rangle \otimes |1, -1\rangle = \sqrt{\frac{1}{2}}|1, 0\rangle - \sqrt{\frac{1}{2}}|0, 0\rangle$$

The four final states are:

(a) $|\pi^+\Sigma^-\rangle = |1, 1\rangle \otimes |1, -1\rangle = \sqrt{\frac{1}{6}}|2, 0\rangle + \sqrt{\frac{1}{2}}|1, 0\rangle + \sqrt{\frac{1}{3}}|0, 0\rangle$
(b) $|\pi^0\Sigma^0\rangle = |1, 0\rangle \otimes |1, 0\rangle = \sqrt{\frac{2}{3}}|2, 0\rangle + 0|1, 0\rangle - \sqrt{\frac{1}{3}}|0, 0\rangle$
(c) $|\pi^-\Sigma^+\rangle = |1, -1\rangle \otimes |1, 1\rangle = \sqrt{\frac{1}{6}}|2, 0\rangle - \sqrt{\frac{1}{2}}|1, 0\rangle + \sqrt{\frac{1}{3}}|0, 0\rangle$
(d) $|\pi^0\Lambda^0\rangle = |1, 0\rangle \otimes |0, 0\rangle = 1|1, 0\rangle$

from which one obtains:

(a) $\langle \pi^+\Sigma^-|A|K^- p\rangle = \frac{1}{2}a_1 - \sqrt{\frac{1}{6}}a_0$
(b) $\langle \pi^0\Sigma^0|A|K^- p\rangle = \sqrt{\frac{1}{6}}a_0$
(c) $\langle \pi^-\Sigma^+|A|K^- p\rangle = -\frac{1}{2}a_1 - \sqrt{\frac{1}{6}}a_0$
(d) $\langle \pi^0\Lambda^0|A|K^- p\rangle = \sqrt{\frac{1}{2}}a_0$

Problem 7.16 The transition probabilities of Eq. (7.3) was obtained in the case of a purely electromagnetic process involving leptons.

Fig. 7.2 Schematic diagram of the ρ formation and decay

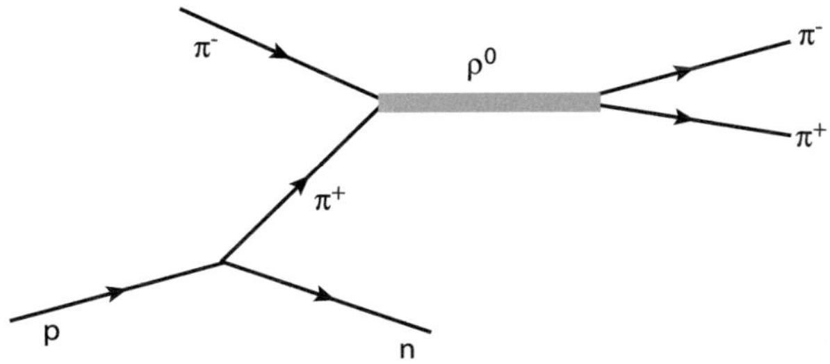

In the case of an annihilation process of neutral vector mesons (Fig. 7.19) $\rho^0 = (u\bar{u} - d\bar{d})/\sqrt{2}$, $\omega^0 = (u\bar{u} + d\bar{d})/\sqrt{2}$, $\phi^0 = s\bar{s}$ (collectively denoted as V^0) made of quark-antiquark pairs with electric charge Q_i in units of the proton electric charge, the partial width Γ is obtained from Eq. (7.3) with the replacements $\alpha \to \alpha Q_i$ and $m_e \to m_V$:

$$\Gamma\left(V^0 \to \ell^+\ell^-\right) = \frac{16\pi\alpha^2(\sum_i Q_i)^2}{m_V^2}|\psi(0)|^2 \tag{7.27}$$

the sum extends over the flavors of the valence quarks of the meson (only the quark s for the ϕ^0 meson). The wave function $\psi(0)$ represents the spatial superposition of the $q\bar{q}$ system. It can be considered as unknown; however, since the interactions between quarks do not depend on the quark flavors, it can be hypothesized that $\psi(0)$ is, in first approximation, identical for the three vector mesons considered.

Problem 7.17 Here, we consider the formation and decay of a light meson resonance; as schematically illustrated in Fig. 7.2, the ρ resonance is formed in $\pi^- p$ collisions and subsequently decays into two $\pi^+\pi^-$ which are pseudoscalar mesons (i.e., with $J^P = 0^-$).

The ρ can be regarded as a bound state of two spin zero pions with orbital angular momentum $\ell = 1$. The final state parity is $P_f = P_{\pi^+} \times P_{\pi^-} \times (-1)^\ell = -1$. For the initial state, one has therefore $J = 1$ and odd parity: $J_\rho^P = 1^-$. Note that the J^P assignment to a resonance such as the ω, which decays into 3π, is more complex and requires the use of a *Dalitz diagram*.

Let us now determine the isospin of the ρ resonance. The ρ decays into two pions, which are bosons and must therefore have a wave function symmetric under their exchange, including the isospin. With J_f =1, the angular momentum wave function is antisymmetric. To construct a symmetric wave function, the isospin states must also be antisymmetric under the $\pi \leftrightarrow \pi$ exchange, and only odd isospin values are possible. Considering in addition the isospin values of the colliding particles ($p\pi^-$), it can be concluded that the ρ resonance has isospin $I_\rho = 1$. Note that indeed, the ρ exists in three different charged states, ρ^-, ρ^0, ρ^+, respectively with the third isospin components $I_3 = -1, 0, 1$. The G-parity being defined as $-1^{\ell+J}$, one has $I_\rho^G = 1^+$.

The ρ meson is a charge eigenstate with eigenvalue $C = -1$. The charge conjugation operation implies the transformation $\pi^+\pi^- \to \pi^-\pi^+$ which corresponds to the inversion of the spatial coordinates, i.e., to the parity transformation: $C|\pi^+\pi^-\rangle = (-1)^\ell|\pi^-\pi^+\rangle = -|\pi^-\pi^+\rangle$.

Therefore, one finally can write for the ρ resonance: $I^G(J^{PC}) = 1^+(1^{--})$.

Note that the spin of the ρ can be measured studying the angular distribution of the flight direction of the two pions in the c.m. system [H91]. It is experimentally verified that $J_\rho = 1$. In fact, the flight direction of the ρ in the c.m. system can be chosen as the z axis. In this case, the ρ is polarized and considering the angular distribution of the two pions in the rest system of the ρ, one has $\ell_z = 0$. The eigenfunction of the two pions contains the spherical harmonic function Y_ℓ^0 which gives rise to an angular distribution of the type $|Y_\ell^0(\theta, \varphi)|^2$; for $\ell = 1$, this distribution reduces to a $\cos^2\theta$ distribution only. In practice, a slight deviation from a $\cos^2\theta$ distribution is observed: the distribution can be more precisely described by the function: $A + B\cos\theta + C\cos^2\theta$. This can be explained in terms of interference of the amplitude of the background, which is spherically symmetric, with the amplitude of the resonance.

The reason why no resonance is observed in the $\pi^0\pi^0$ channel is explained in Problem 7.6.

Chapter 8
Weak Interactions and Neutrinos

Problems

8.1. **Universality of weak interactions**. According to the Puppi triangle (see Sect. 8.4.3), the weak interaction occurs with the same characteristics and same intensity for beta nuclear decays, for the muon decay and for the nuclear capture of negative muons. Explain the reason, using Feynman diagrams.
[See solutions]

8.2. **Neutron and muon decay**. Taking into account that the neutron lifetime is $\tau_n = 887$ s, and that of the muon $\tau_\mu = 2.2 \cdot 10^{-6}$ s, show that the couplings in these two cases have the same order of magnitude when considering the phase space factor.
[Hint: See Sect. 8.4.1]

8.3. **Feynman diagrams**. Draw the Feynman diagrams for the following decay and interaction processes:
(a) $\mu^- \to e^- \overline{\nu}_e \nu_\mu$;
(b) $\tau^+ \to e^+ \nu_e \overline{\nu}_\tau$;
(c) $\nu_\mu e^- \to \nu_\mu e^-$;
(d) $\mu^+ e^- \to \overline{\nu}_\mu \nu_e$.
[See solutions]

8.4. **Axial and vector couplings**. Using the data for the β decay of ${}^{14}_{8}\mathrm{O}$ and of ${}^{6}_{2}\mathrm{He}$ reported in Table 8.1, determine the ratio $|g_A/g_V|$.
[See solutions]

S. Braibant et al., *Particles and Fundamental Interactions: Supplements, Problems and Solutions*, Undergraduate Lecture Notes in Physics,
DOI 10.1007/978-94-007-4135-5_8,

8.5. **Sargent rule from $\Sigma^{\pm}$ decays**. The Sargent rule, Eq. (8.18):

$$W = \frac{(\Gamma_i/\Gamma)}{\tau} \simeq G_F^2 E_0^5 \simeq G_F^2 \Delta m^5 \tag{8.1}$$

can be tested through the $\Sigma^{\pm}$ decay.

(a) Show in terms of quark content that the Σ^+ is not the antiparticle of the Σ^-.

(b) Draw the Feynman diagrams and calculate, using the particle masses reported in the *Review of Particle Physics* [P10], the ratio between the lifetimes of the following semileptonic decays:

$$\Sigma^+ \to \Lambda^0 e^+ \nu_e; \qquad \Sigma^- \to \Lambda^0 e^- \overline{\nu}_e$$

(c) Using the lifetimes and branching ratios reported in [P10] for the above semileptonic decays, compare the measured ratio of decay fractions with the expected one.

[See solutions]

8.6. **Strong and weak interaction lifetimes: the ρ^0 and K^0 decays**. The ρ^0 and K^0 mesons decay mainly in $\pi^+\pi^-$. Explain why the ρ^0 lifetime is of the order of 10^{-23} s and that of the K^0 of the order of 10^{-10} s. Draw the Feynman diagrams for both decays.

[See solutions]

8.7. **Pion decay branching ratios**. Calculate the ratio of the phase space factors for the following decays:

$$\pi^- \to e^- \overline{\nu}_e$$

$$\pi^- \to \mu^- \overline{\nu}_\mu$$

Estimate the two lifetimes and compare them with the experimental values. Determine if the results are consistent with the hypothesis that the lifetime is purely determined by the phase space factor.

[Solution: see Sect. 8.10]

8.8. **Strange and charmed particle decay**. Draw the Feynman diagrams, specify the couplings and comment the following decays (in bracket, the composition in terms of the valence quarks):

(a) $\Lambda_c^+ \quad [cud] \to \pi^+ \Lambda^0$

(b) $D^+ \quad [c\overline{d}] \to \pi^+ \overline{K}^0$

(c) $B^+ \quad [\overline{b}u] \to \pi^+ \overline{D}^0$

(d) $\overline{B}_s^0 \quad [b\overline{s}] \to \pi^- D_s^+$

(e) $\Lambda_b^0 \quad [bud] \to \pi^- \Lambda_c^+$.

[See solutions]

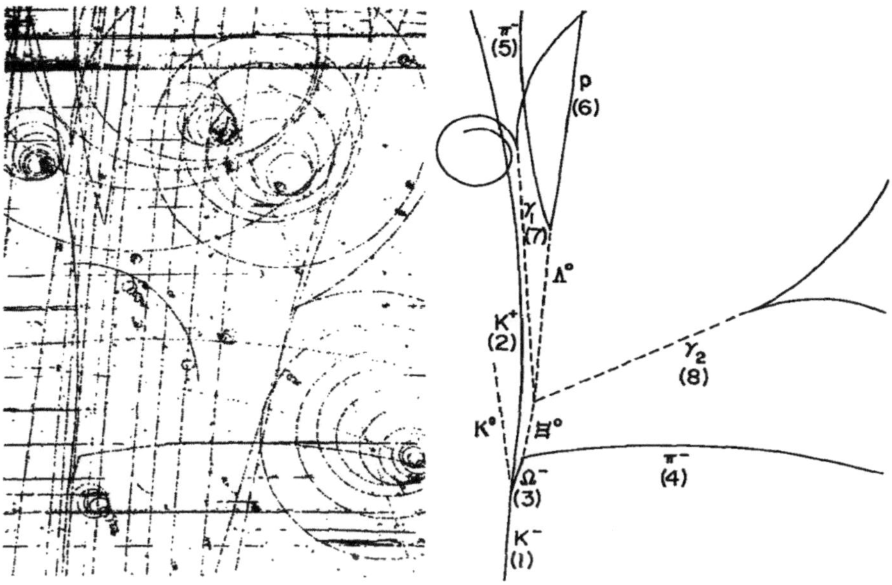

Fig. 8.1 Bubble chamber trace of the first observed Ω^- baryon [8B64]. The event was discovered in 1964 by a team of physicists from the Brookhaven National Laboratory, the University of Rochester and Syracuse University, led by N. Samios of Brookhaven, using the 80-inch bubble chamber

8.9. **Non-leptonic D^0 decays**. Draw the Feynman diagrams of the following D^0 meson decays and estimate the relative amplitudes:
(a) $D^0 \to K^-\pi^+$
(b) $D^0 \to \pi^-\pi^+$
(c) $D^0 \to K^+\pi^-$
[See solutions]

8.10. **Suppression of $\Delta S = 1$ NC interaction**. It is experimentally found that the NC/CC ratio for the charged K decays is:

$$\left(K^+ \to \pi^+\nu\overline{\nu}\right)/\left(K^+ \to \pi^0\mu^+\nu_\mu\right) < 10^{-8}$$

This is one of the experimental evidence that the weak neutral current decays with $\Delta S = 1$ change in strangeness are suppressed.
(a) Draw the Feynman diagrams of the two decays.
(b) Check that, assuming the existence of the c quark, the transition probabilities induced by a $\Delta S = 1$ neutral current is vanishing.
[See solutions]

8.11. **The Ω^- decay products**. Figure 8.1 shows the bubble chamber picture of the first produced Ω^- event. An incoming K^- meson interacts with a proton in the liquid hydrogen of the bubble chamber and produces an Ω^-, a K^0 and a

K^+ meson. All these unstable hadrons decay into other particles. To have an idea of the scale of the picture, the length of the Ξ^0 track is 3 cm.

(a) Show that the reaction $K^- p \to \Omega^- K^0 K^+$ is the minimal reaction for the production of a Ω^-;

(b) Evaluate the minimum momentum of the K^- in the laboratory system in order to produce the Ω^-. The K^- of the experimental beam had a momentum of 5 GeV/c.

(c) Discuss the decay of the Ω^- presented in the picture.

[See solutions]

8.12. **Feynman diagrams and couplings**. Draw the Feynman diagrams, specify the couplings and comment the following decays:

(a) $\Lambda^0 \to p e^- \overline{\nu}_e$

(b) $\Xi^- \to \Lambda^0 \pi^-$

(c) $\nu_\mu p \to \mu^- \Delta^{++}$

(d) $D^0 \to K^- \mu^+ \nu_\mu$

(e) $D^0 \to K^+ \mu^- \overline{\nu}_\mu$.

[See solutions]

8.13. **Muon and tau decay**. The muon is a charged particle whose mass is 105 MeV/c^2 and lifetime $\sim 2 \cdot 10^{-6}$ s. It decays into an electron ($m_e =$ 0.5 MeV/c^2), a neutrino and an antineutrino.

(a) The electron is the only observable particle in the muon decay. Explain the characteristic of the electron spectrum necessary to demonstrate that the muon decay at rest is not a two-body decay.

(b) In the case of muon decaying at rest, calculate the maximum electron energy if three particles are present in the final state.

(c) The τ ($m_\tau = 1777$ MeV/c^2) lepton decays with the same characteristic of the muon. Estimate the τ lifetime.

8.14. **The CKM matrix**. Using the CKM matrix (data from [P10]):

$$\begin{pmatrix} V_{ud} & V_{us} & V_{ub} \\ V_{cd} & V_{cs} & V_{cb} \\ V_{td} & V_{ts} & V_{tb} \end{pmatrix}$$

$$= \begin{pmatrix} 0.97428 \pm 0.00015 & 0.2253 \pm 0.0007 & 0.00347 \pm 0.00016 \\ 0.2252 \pm 0.0007 & 0.97345 \pm 0.00015 & 0.041 \pm 0.001 \\ 0.0086 \pm 0.0003 & 0.040 \pm 0.001 & 0.99915 \pm 0.00005 \end{pmatrix}$$

calculate the decay fraction (or branching ratios, BR) of the W boson decays into all possible quark-antiquark and lepton-antilepton pairs. Remember that the sum of all the BRs must be equal to 1. For the hadronic decays, the color factor $N_c = 3$ must be used.

[See solutions]

8.15. **Bilinear forms**. Regarding the bilinear forms defined in Sect. 8.16.1, prove that:
(a) The scalar $\overline{\psi}\psi$ is a relativistic invariant quantity.
(b) The four-vector current $\overline{\psi}\gamma^{\mu}\psi$ is a relativistic invariant quantity.
[See solutions]

8.16. **Phase space in neutron decay**. Demonstrate that in the case of the neutron decay, and neglecting the mass of the electron in the final state, the integral of the phase space Eq. (8.9) gives:

$$\int_0^{E_0/c} p_e^2(E_0 - E_e)^2 dp_e = \frac{E_0^5}{30c^3}. \tag{8.2}$$

[See solutions]

8.17. **Tau decay branching ratios**. The decay fractions of the τ^- decay are reported below (data from [P10]):

Particle	Mass (MeV)	Mean life $\times 10^{-15}$ (s)	Decay mode	Decay fraction (Γ_i/Γ)
τ^-	1776.82 ± 0.16	(290.6 ± 1.0)	$\mu^-\overline{\nu}_\mu\nu_\tau$	$(17.36 \pm 0.05)\%$
			$e^-\overline{\nu}_e\nu_\tau$	$(17.85 \pm 0.05)\%$
			$h^-\nu_\tau$	$(11.61 \pm 0.06)\%$
			$\pi^-\nu_\tau$	$(10.91 \pm 0.07)\%$
			$K^-\nu_\tau$	$(6.96 \pm 0.23) \times 10^{-3}$
			$h^- \geq 1$ neutral ν_τ	$(37.06 \pm 0.10)\%$

(a) Explain why the branching ratios for $\tau^- \to \mu^-\overline{\nu}_\mu\nu_\tau$ is almost equal to that of $\tau^- \to e^-\overline{\nu}_\mu\nu_\tau$.
(b) Evaluate the expected ratio between decay fractions

$$\frac{\Gamma(\tau^- \to hadrons\ \nu_\tau)}{\Gamma(\tau^- \to \mu^-\overline{\nu}_\mu\nu_\tau)} \tag{8.3}$$

and compare the prediction to the measured value.
(c) Compare the measured ratio between decay fractions:

$$\frac{\Gamma(\tau^- \to K^-\nu_\tau)}{\Gamma(\tau^- \to \pi^-\nu_\tau)} \tag{8.4}$$

with the expected one.
[See solutions]

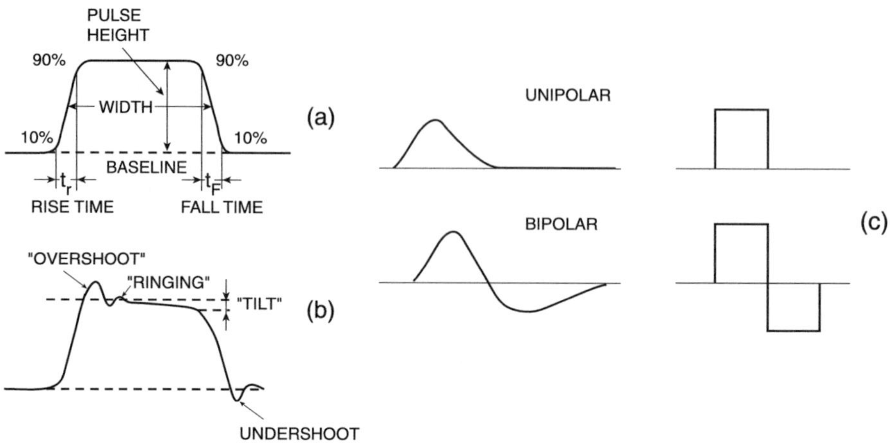

Fig. 8.2 Pulse signal shape: (**a**) almost rectangular and (**b**) less rectangular. (**c**) Unipolar and bipolar signals. The *horizontal axis* measures time while the *vertical axis* indicates an electric current or voltage [L87]

8.18. **$\Delta S = 1$ K decay**. Draw a possible Feynman diagram for the $K^0 \to \mu^+\mu^-$ decay. What can we conclude from the fact that the measured branching ratio of this reaction is $< 10^{-7}$?
[See solutions]

8.19. **WI decay with spectator quarks**. Show that in the hadron model in which a quark undergoes a decay and the others act as "spectators", the lifetimes of the $D^+(c\overline{d})$, $D^0(c\overline{u})$ and $D_s^+(c\overline{s})$ mesons are almost equal.
[See solutions]

Supplement 8.1: Signals, Data Transmission and Electronics

Particle detectors provide information in the form of electric signals, which must be processed by a series of electronic components followed by on-line and off-line computers. Here, a few examples of these rather complex systems are schematically illustrated, emphasizing their modularity, i.e., the possibility to combine them into modules that can be exchanged. More details can be found in specialized texts [L87, 8T10].

Electric Signal Figure 8.2 shows an *output signal* of a particle detector. Its main characteristics are:

(i) *Baseline*: it is the base reference voltage or current level in absence of signal.
(ii) *Signal height* or *amplitude:* the amplitude is the signal maximum height with respect to the baseline.

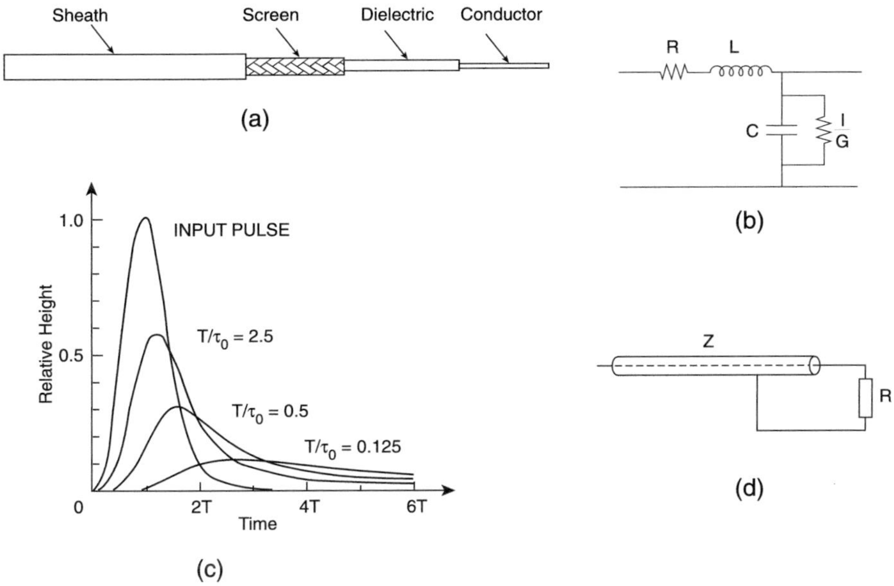

Fig. 8.3 (**a**) Schematic of a coaxial cable and (**b**) its equivalent circuit per unit length, (**c**) distortion of the signal, (**d**) cable termination with a terminating resistor R

(iii) *Signal width*: the signal width is the width at half of the signal maximum.
(iv) *Rise time*: the rise time refers to the time, measured on the initial part of the pulse, required for the signal to change from 10% to 90% of its amplitude.
(v) *Fall time*: the fall time refers to the time, measured on the final part of the pulse, required for the signal to change from 90% to 10% of its amplitude.
(vi) *Analog signals* such as the signals shown in Fig. 8.2: they are continuous signals containing the whole pulse information (such as amplitude, shape, etc.) as a function of time.
(vii) *Digital or logic signals*: they are quantized in a sequence of discrete values, usually: *yes* or *no*.
(viii) The signals are *fast* if the rise time is typically of the order of a few nanoseconds.
(ix) The signals are *slow* if the rise time is typically of the order or larger than a μs.
(x) *Bandwidth*: The Fourier analysis of a signal provides information about the frequency and signal distortion. An electronic system sustains the signal linearly, i.e., without distortion, in a certain frequency bandwidth, while outside of the bandwidth, some frequencies are attenuated or eliminated. A fast electronics, used with large signals about $\Delta t = 5$ ns wide, must have a bandwidth $\Delta\nu \geq 1/\Delta t \simeq 200$ MHz.

Electric Signal Transmission Electrical signals are usually transmitted through coaxial cables having a characteristic impedance (see Fig. 8.3a):

(i) *characteristic impedance* (typical): $Z = V/I \simeq \sqrt{L/C}$, Fig. 8.3b.
(ii) *Termination*: a cable must be terminated on an impedance equal to its typical impedance, Fig. 8.3d.
(iii) *Reflections*: if a cable is not terminated with a resistance equal to its characteristics impedance, a signal is generated at the cable end and is reflected with the same sign, if the final impedance is high, with the opposite sign if the impedance is small.
(iv) *Signal distortion*: in passing through a cable, the signal is distorted as shown in Fig. 8.3c.

Electronics Consider the following modules:

(i) *Preamplifiers*: they are usually very close to the detector; they amplify weak signals and send them on a cable to the rest of the electronics. They amplify voltages, charges or currents.
(ii) *Amplifiers*: they amplify signals from a preamplifier and form them in a way to be conveniently used later.
(iii) *Non-linear amplifiers*: they can amply, for instance, the part of the signal above a given threshold.
(iv) *Shapers*: they shape the signals through delay lines or RC differentiator and integrator circuits.
(v) *Fan-out*: are active or passive circuits that distribute a signal to various electronic circuits.
(vi) *Fan–in*: they accept different signals and distribute a signal sum.
(vii) *Discriminators*: they respond to input signals with a height above a certain threshold; they produce a standard digital output.
(viii) *Differential discriminators* (single channel analyzer): they are characterized by two thresholds: a lower and a higher; they allow to select only the signal pulses whose amplitudes lie within the two thresholds.
(ix) *Analog-to-Digital Converters (ADC)*: they convert the height or the integral of a signal into a digital number. For instance, the height of a signal between 0 and 5 V may be converted by a 10-bit ADC in a number between 0 and $2^{10} - 1 = 1023$. The time necessary for the ADC to perform the analog-to-digital conversion is ≥ 1 μs. The so-called *flash ADC* are very fast compared to other ADC types, so a single ADC flash can be used to analyze various channels in sequence, or to analyze the subsequent amplitudes of a pulse, functioning in this way as a *Waveform Analyzer (WFA)*.
(x) *Multi-Channel Analyzers (MCA)*: they sort the incoming pulses according to their height and record them in dedicated memories. An ADC may be an essential element of a MCA.
(xi) *Time-to-Digital Converters (TDC)*: they convert the time between two pulses into a digital number.
(xii) *Coincidence circuits*: these are logical units with two input channels that provide a logic output signal when two input signals are present at "the same time", i.e., when the two input signal are coincident.

(xiii) *Flip–flop*: this is a circuit with two logic inputs that remains stable in one of the two logic states until the arrival of a control signal which makes it change state.
(xiv) *Gate generators*: they are controllable unit generating a signal of appropriate length that may be used as a "gate" for ADC, TDC, analyzers, etc....
(xv) *Recorders*: they record and store signal pulses.
(xvi) *Ratemeter*: it instantly provides the average number of pulses arriving in a certain unit of time, for example, every 2 s.
(xvii) *Scaler*: this is the unit that counts the number of pulses and displays it in a viewer.
(xviii) *Attenuators*: in some cases, it is necessary to attenuate the signal pulse before processing it.

Some electronic modules can be combined to form the *electronic logic* of an experiment (or of a part of it). The electronic logic is mainly used to define simple criteria to rapidly decide if an event must be permanently stored (the *trigger*, see Supplement 9.1). Trigger systems are necessary when only a small fraction of the total event rate can be recorded due to real-world limitations in data storage capacity and rates. Since experiments are typically searching for *interesting* events (such as decays of rare particles, or the combination of the positron and neutron signals, as in the Cowan and Reines neutrino experiment [8C97]) that occur at a relatively low rate, trigger systems are used to identify the events that should be recorded for later analysis.

The present generation of experiments at accelerators can have event rates larger than 1 MHz and trigger rates that can be below 10 Hz. For example, the Large Hadron Collider (LHC) has a huge event rate and each experiment has up to hundred million electronic channels. For this reason (as shown in Supplement 10.1), data must be reduced by an *hardware trigger*, a *software trigger*, and processed with huge computer facilities.

Solutions

Problem 8.1 All processes foresee the exchange of a $W^{\pm}$ boson. The muon decay (for instance, $\mu^- \to e^- \overline{\nu}_e \nu_\mu$) is shown in Problem 8.3a. A $\Delta S = 0$ process corresponds, e.g., to the neutron decay, as shown in Fig. 8.3. The muon capture corresponds to the process: $\mu^- p \to n \nu_\mu$, and its Feynman diagram is equal to that shown in Fig. 5.5c with the exchange $e^- \to \mu^-$, $\nu_e \to \nu_\mu$.

Problem 8.3 See Figs. 8.4, 8.5, 8.6, 8.7.

Problem 8.4 The decay ${}^{14}_{8}\mathrm{O} \to {}^{14}_{7}\mathrm{N}^* \ e^+ \nu$ is a Fermi transition; the transition probability, given in Eq. (8.3), is proportional to:

$$G_F^2 |\mathcal{M}_F|^2 = G_F^2 m_{i,s}^{V\,2} g_V^2 = 1.52 \times 10^{-8}\ \mathrm{MeV}^2\,\mathrm{fm}^6$$

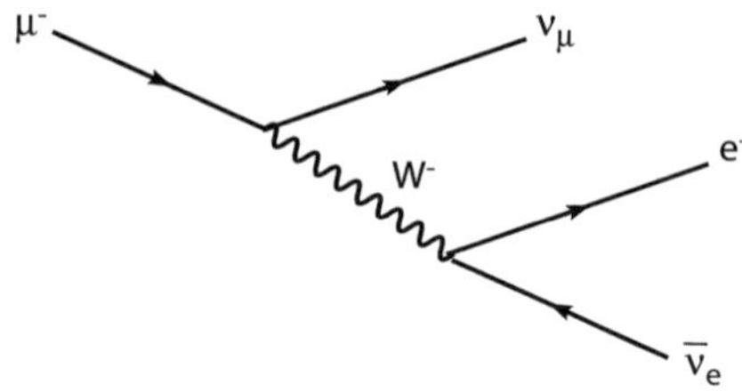

Fig. 8.4 Problem 8.3(a) Feynman diagram for $\mu^- \to e^- \overline{\nu}_e \nu_\mu$

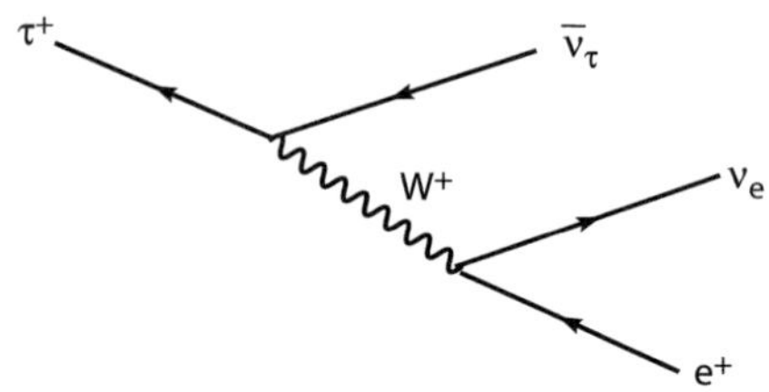

Fig. 8.5 Problem 8.3(b) Feynman diagram for $\tau^+ \to e^+ \nu_e \overline{\nu}_\tau$

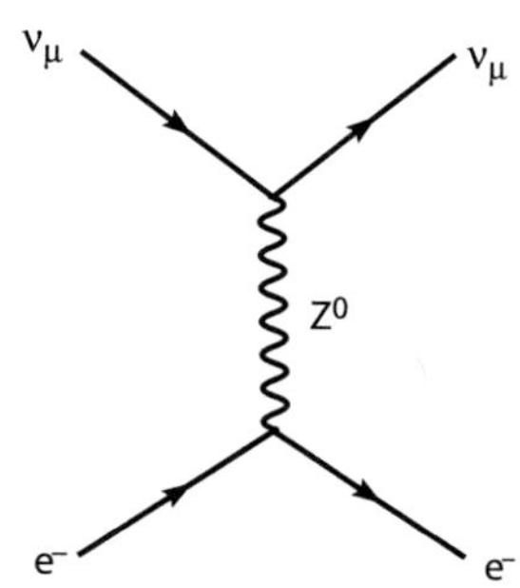

Fig. 8.6 Problem 8.3(c) Feynman diagram for $\nu_\mu e^- \to \nu_\mu e^-$. A similar diagram holds for the $\nu_\tau e^- \to \nu_\tau e^-$ scattering. For the $\nu_e e^- \to \nu_e e^-$ scattering, in addition to the Z^0, there is also the $W^\pm$ exchange, as shown in Fig. 12.10

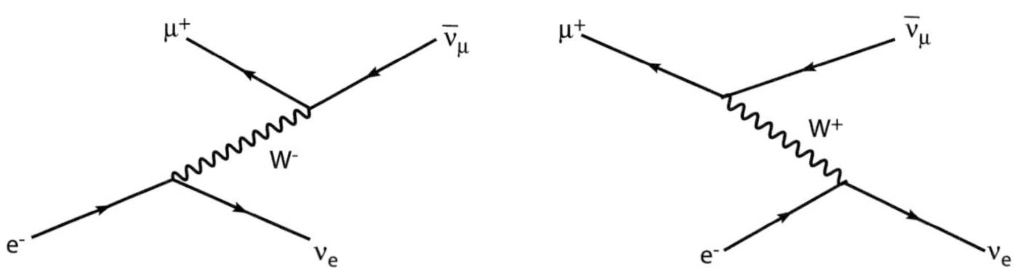

Fig. 8.7 Problem 8.3(d) Feynman diagrams for $\mu^+ e^- \to \overline{\nu}_\mu \nu_e$. As explained in Fig. 4.1 for an electromagnetic scattering, the process of a W^- emission from the electron (*left*) and subsequent absorption by the μ^+ is indistinguishable from the process of a W^+ emission from the μ^+ and absorption from the e^- (*right*)

where the spin and isospin multiplicity factor is $m^V_{i,s} = 2$ (see Table 8.1).

The decay ${}^6_2\text{He} \to {}^6_3\text{Li}\ e^-\overline{\nu}$ occurs through a pure Gamow-Teller transition (the change in the nuclear angular momentum is equal to one). In this case, the factor

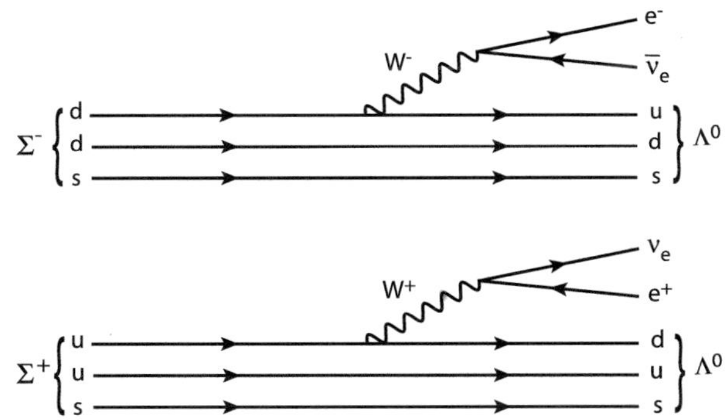

Fig. 8.8 $\Sigma^{\pm}$ composition in term of valence quarks, and their semileptonic decay

$m^A_{i,s} = 6$ and one has:

$$G_F^2|\mathcal{M}_{GT}|^2 = G_F^2 m^A_{i,s} g_A^2 = 7.45 \times 10^{-8}\ \text{MeV}^2\,\text{fm}^6$$

Calculating the ratio between the two transition probabilities, one obtains:

$$\left(\frac{6g_A^2}{2g_V^2}\right) = \left(\frac{7.45 \times 10^{-8}}{1.52 \times 10^{-8}}\right)$$

from which, one derives that:

$$\left|\frac{g_A}{g_V}\right| = 1.27$$

value which is very close to the value reported in Eq. (8.26).

Problem 8.5

(a) In term of quark composition, the baryons with strangeness $S = 1$ are:

$$\Sigma^+ = (uus), \qquad m_{\Sigma^+} = 1189.37\ \text{MeV};$$
$$\Sigma^- = (dds), \qquad m_{\Sigma^-} = 1197.45\ \text{MeV}$$

Because of the slight mass difference between the u, d quarks, there is a mass difference between the positive and negative Σ baryons.

(b) The Feynman diagrams for the Σ^+ and Σ^- semileptonic decay into a Λ^0 are shown in Fig. 8.8. According to the Sargent rule, Eq. (8.1), the probability transition for a three body decay depends on the fifth power of the free energy available in the final state E_0, and $E_0 \simeq \Delta m$, the mass difference between hadrons in the initial and final states. The final state particle masses, in the considered semileptonic Σ^+ and Σ^- decays, are the same, with $m_\Lambda = 1115.68$ MeV. Therefore, one obtains:

$$\text{in the } \Sigma^+ \to \Lambda^0 e^+ \nu_e: \quad \Delta m_+ = (1189.37 - 1115.68) = 73.7\ \text{MeV}$$
$$\text{in the } \Sigma^- \to \Lambda^0 e^- \overline{\nu}_e: \quad \Delta m_- = (1195.45 - 1115.68) = 79.8\ \text{MeV}$$

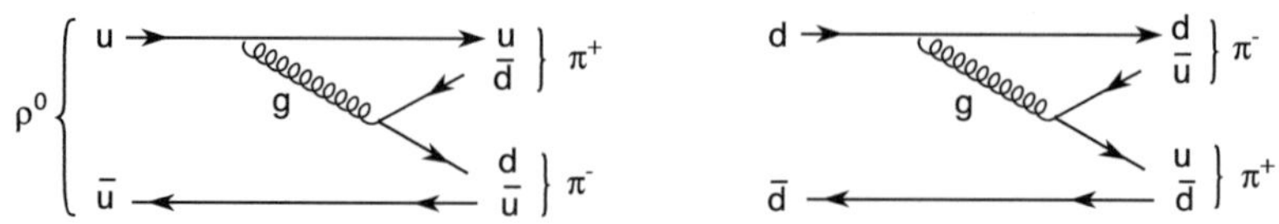

Fig. 8.9 Strong interaction decay $\rho^0 \to \pi^+\pi^-$

Fig. 8.10 Weak interaction decay $K^0 \to \pi^+\pi^-$

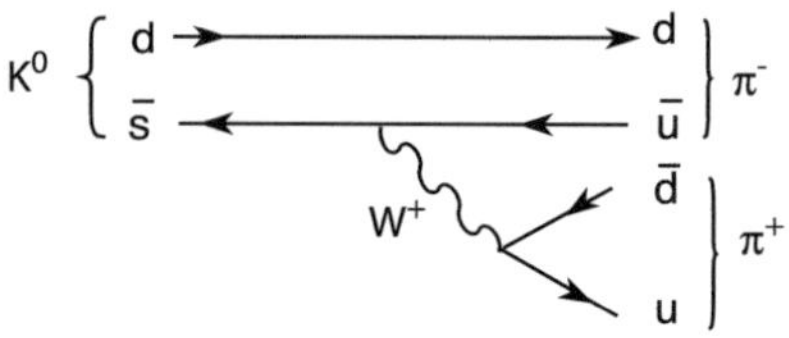

The predicted lifetime ratio, according to the Sargent rule, is:

$$\frac{\tau_{\Sigma^-}}{\tau_{\Sigma^+}} = \frac{W_+}{W_-} \simeq \left(\frac{73.7}{79.8}\right)^5 = 0.67 \tag{8.5}$$

(c) The measured ratio is obtained from the $\Sigma^\pm$ lifetimes and branching ratios:

$$W_+ = BR_+ \cdot \frac{1}{\tau_{\Sigma^+}}; \qquad W_- = BR_- \cdot \frac{1}{\tau_{\Sigma^-}}$$

where (from PDG):

$$BR_+ = BR(\Sigma^+ \to \Lambda^0 e^+ \nu_e) = 2.0 \times 10^{-5}$$
$$BR_- = BR(\Sigma^- \to \Lambda^0 e^- \overline{\nu}_e) = 5.7 \times 10^{-5}$$

$\tau_{\Sigma^+} = 0.8018 \cdot 10^{-10}$ s and $\tau_{\Sigma^-} = 1.479 \cdot 10^{-10}$ s. Finally, one obtains:

$$\left(\frac{W_+}{W_-}\right)_{mea} = \frac{BR_+ \cdot \tau_{\Sigma^-}}{BR_- \cdot \tau_{\Sigma^+}} = \frac{2.0 \cdot 10^{-5} \cdot 1.479 \cdot 10^{-10}}{5.7 \cdot 10^{-5} \cdot 0.8018 \cdot 10^{-10}} \simeq 0.64 \tag{8.6}$$

which is in good agreement with the expected value (8.5).

Problem 8.6 The ρ^0 decay in two pions is due to the strong interaction (SI), Fig. 8.9. The corresponding lifetime is that characteristic of the SI. In terms of quarks, the ρ composition is $\rho^0 = \frac{1}{\sqrt{2}}(u\bar{u} - d\bar{d})$.

In terms of quarks, the K^0 composition is $K^0 = d\bar{s}$; it is the lighter strange meson and cannot decay through the strong interaction, which conserves the strangeness. It can therefore decay only through the weak interaction with a much longer lifetime, Fig. 8.10.

Problem 8.8 In the first two decays shown here below, there is a transition from a quark c to a quark s implying a strangeness variation ($\Delta S = 1$). These decays cannot proceed via strong interaction, which conserves the quark flavor, but can arise only through weak interaction, with much longer lifetimes. For the last three decays, one has instead a transition with a flavor variation from a quark b to a quark c.

(a) In terms of quarks, $\Lambda_c^+ = [cud]$; it is a charmed baryon. The considered decay is $\Lambda_c^+ \to \pi^+ \Lambda^0$. The Λ^0 composition is $\Lambda^0 = [sud]$.

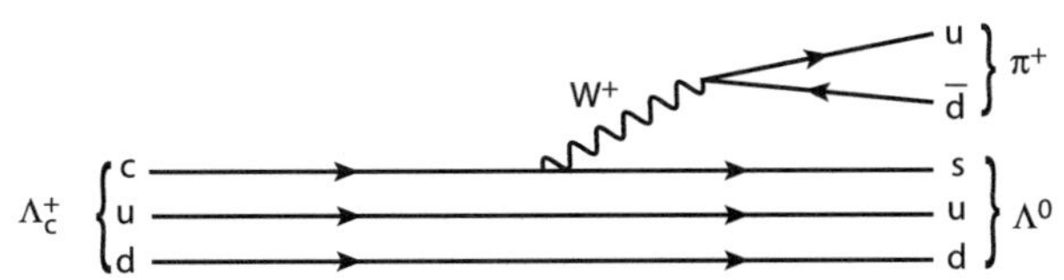

(b) The D^+ composition is $D^+ = [c\overline{d}]$; it is a charmed meson. The decay considered here is $D^+ \to \pi^+ \overline{K}^0$. The $\overline{K}^0$ composition is $\overline{K}^0 = [s\overline{d}]$.

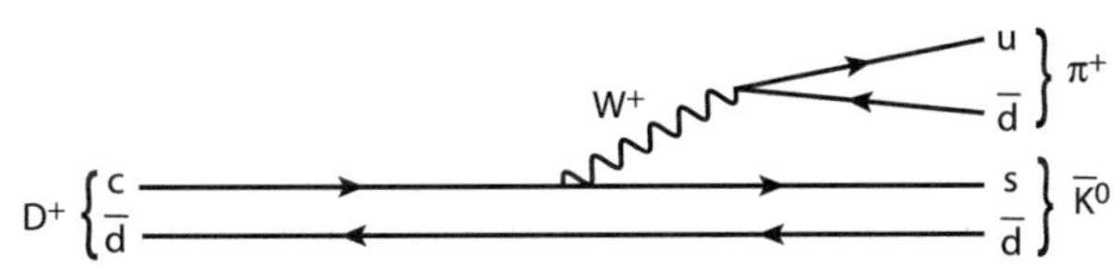

(c) The B^+ composition is $B^+ = [\overline{b}u]$; it is a beauty meson. The decay designed here is $B^+ \to \pi^+ \overline{D}^0$. The $\overline{D}^0$ composition is $\overline{D}^0 = [\overline{c}u]$.

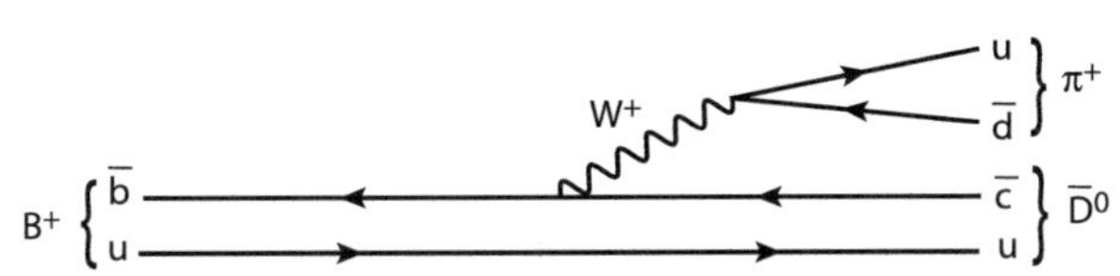

(d) The $\overline{B_s^0}$ composition is $\overline{B_s^0} = [b\overline{s}]$; it is a beauty meson. The decay designed here is $\overline{B_s^0} \to \pi^- D_s^+$. The D_s^+ composition is $D_s^+ = [c\overline{s}]$.

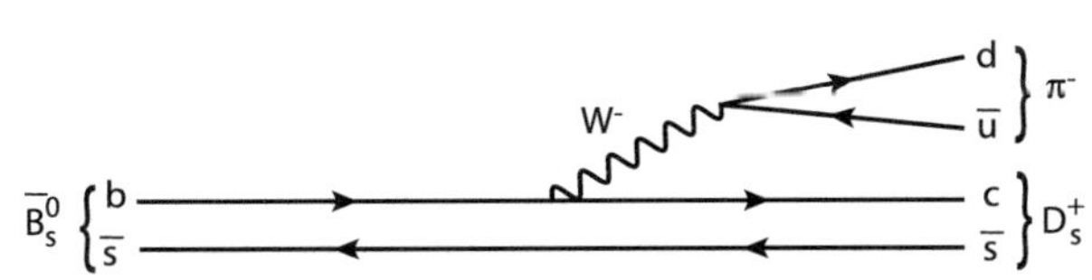

(e) The Λ_b^0 composition is $\Lambda_b^0 = [bud]$; it is a beauty baryon. The decay designed here is $\Lambda_b^0 \to \pi^- \Lambda_c^+$. The Λ_c^+ composition is $\Lambda_c^+ = [cud]$.

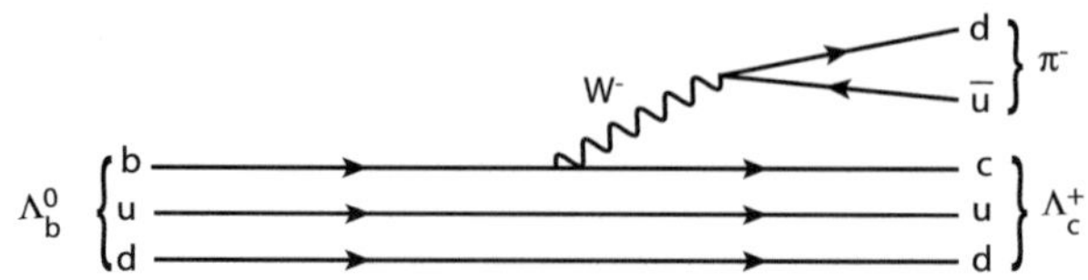

Problem 8.9 They are three non-leptonic weak decays. In terms of valence quarks, the compositions of the particles involved are:

$$D^0 = [c\bar{u}], \qquad K^- = [s\bar{u}], \qquad K^+ = [\bar{s}u],$$
$$\pi^+ = [u\bar{d}], \qquad \pi^- = [d\bar{u}]$$

(a) $D^0 \to K^-\pi^+$
It is a *Cabibbo favored decay*: the coupling constants at the two vertices connected by the W vector boson correspond to transitions of the type $c \to s$ and $u \to d$, respectively. Both amplitudes are proportional to $\cos^2\theta_C$. The transition probability (the product of the two amplitudes) is proportional to $[\cos^2\theta_C]^2 \simeq 0.90$.

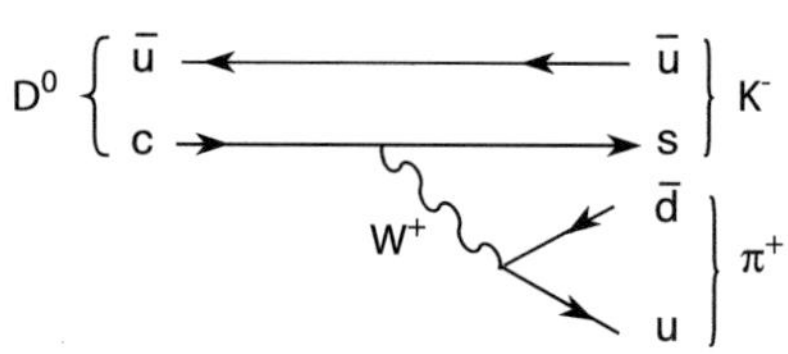

(b) $D^0 \to \pi^-\pi^+$
It is a *Cabibbo suppressed decay*: the coupling constant corresponding to the $c \to d$ transition is proportional to $\sin\theta_C$; the coupling constant corresponding to the $u \to d$ transition is proportional to $\cos\theta_C$. The total transition probability is proportional to the product of the square of the coupling constant: $\sin^2\theta_C \cos^2\theta_C \simeq (0.22)^2 \simeq 0.05$.

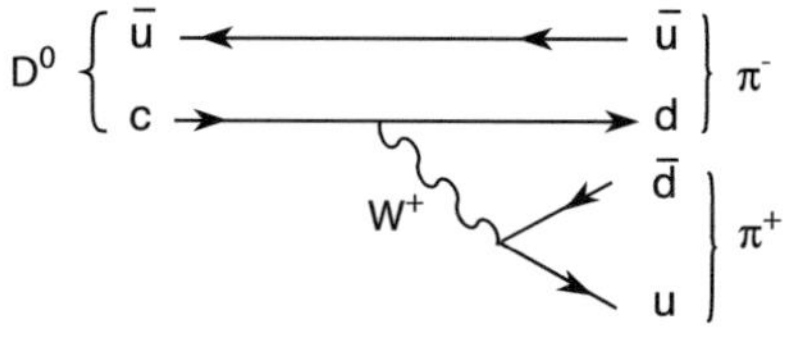

(c) $D^0 \to K^+\pi^-$
It is a *Cabibbo doubly suppressed decay*: the coupling constant corresponding to the $c \to d$ vertex is proportional to $\sin\theta_C$; the coupling constant corresponding to the $s \to u$ vertex is also proportional to $\sin\theta_C$. The total transition probability

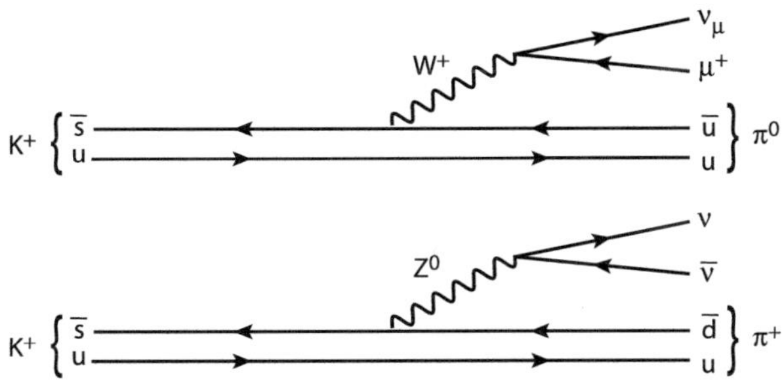

Fig. 8.11 Feynman diagrams for the charged current (CC) weak decay $K^+ \to \pi^0 \mu^+ \nu_\mu$ and the neutral current (NC) weak decay $K^+ \to \pi^+ \nu\overline{\nu}$

is proportional to the product of the square of the coupling constant: $[\sin^2\theta_C]^2 \simeq 3 \times 10^{-3}$.

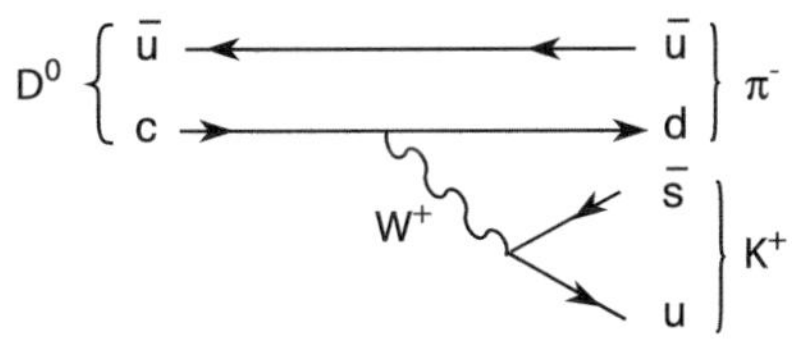

The three processes have therefore relative amplitudes (neglecting mass differences in the final state) (a) : (b) : (c) = 0.90 : 0.05 : 0.003. The measured branching ratios (see the *Review of Particle Physics* [P10]) are (a) : (b) : (c) = 3.8% : 0.14% : 0.015%, in qualitative agreement with the prediction.

Problem 8.10 The Feynman diagrams for the charged current and neutral current K^+ decays are shown in Fig. 8.11.

(b) Assuming the existence of the u, d, s quarks only, the mechanism which rotates the (d, s) quarks: $\binom{u}{d_c} = \binom{u}{d\cos\theta_c + s\sin\theta_c}$ includes a $\Delta S = 1$ term in the NC current with a coupling $\propto \sin\theta_c \cos\theta_c$:

$$(\overline{u}, \overline{d}_c)\begin{pmatrix} u \\ d_c \end{pmatrix} = (\overline{u}, \overline{d}\cos\theta_c + \overline{s}\sin\theta_c)\begin{pmatrix} u \\ d\cos\theta_c + s\sin\theta_c \end{pmatrix}$$
$$= \underbrace{u\overline{u} + (d\overline{d}\cos^2\theta_c + s\overline{s}\sin^2\theta_c)}_{\Delta S=0} + \underbrace{(s\overline{d} + sd)\sin\theta_c\cos\theta_c}_{\Delta S=1}.$$

With the introduction of the c quark and a second doublet: $\binom{c}{s_c} = \binom{c}{s\cos\theta_c - d\sin\theta_c}$, the neutral current can be rewritten in the form:

$$(\overline{u}, \overline{d}_c)\begin{pmatrix} u \\ d_c \end{pmatrix} + (\overline{c}, \overline{s}_c)\begin{pmatrix} c \\ s_c \end{pmatrix}$$
$$= \underbrace{u\overline{u} + c\overline{c} + (d\overline{d} + s\overline{s})\cos^2\theta_c + (s\overline{s} + d\overline{d})\sin^2\theta_c}_{\Delta S=0}$$
$$+ \underbrace{(s\overline{d} + \overline{s}d - \overline{s}d - s\overline{d})\sin\theta_c\cos\theta_c}_{\Delta S=1} \tag{8.7}$$

and the term with $\Delta S = 1$ is automatically cancelled.

Problem 8.11

(a) In terms of quarks, the compositions of particles involved are:

$$K^- = [s\overline{u}]; \qquad p = [uud]; \qquad \Omega^- = [sss]; \qquad K^0 = [d\overline{s}]; \qquad K^+ = [u\overline{s}]$$

The creation of the Ω^- proceeds through the strong interaction; a $u\overline{u}$ pair in the initial state annihilates, and two $s\overline{s}$ pairs are created in the final state:

$$K^- p = [s\overline{u}][uud] \to s(\overline{u}u)ud \to s(s\overline{s}s\overline{s})ud \to [sss][\overline{s}d][\overline{s}u] = \Omega^- K^0 K^+$$

The kaons are the strange mesons with the lowest mass.

(b) The masses of the involved particles are:

$$m_{K^\pm} = 494 \text{ MeV}; \qquad m_{K^0} = 498 \text{ MeV}; \qquad m_p = 938 \text{ MeV};$$
$$m_{\Omega^-} = 1674 \text{ MeV}$$

Using the discussion of Problem 3.10, one has $s = (m_{\Omega^-} + m_{K^0} + m_{K^+})^2 = (2666)^2 \text{ MeV}^2$ and the minimum energy of the incoming K^- is:

$$E_{K^-} = \frac{s - (m_p^2 + m_{K^-}^2)}{2m_p} = 2695 \text{ MeV};$$
$$p_{K^-} = \sqrt{E_{K^-}^2 - m_{K^-}^2} = 2650 \text{ MeV/c}$$

The 5 GeV/c of the beam is sufficient to produce the Ω^- baryon.

(c) The momentum of the Ω^- in the final system is comparable with its mass; the ratio $p/m \sim 1$ and the particle travels, during its lifetime τ, a path length $L = c\tau$ (see Problem 1.9). The Ω^- path length observed in the picture of Fig. 8.1 is $L \sim 2$ cm. Its lifetime is $\tau_\Omega \simeq L/c \sim (2 \text{ cm})/(30 \text{ cm/ns}) \sim 10^{-10}$ s. (Compare the result with the measured lifetime as reported in Table 7.3.)

The Ω^- is a $S = -3$ particle. In the picture, it decays into a neutral particle plus a π^-. The latter is identified by energy-momentum measurement in presence of a magnetic field. As the strong and electromagnetic interactions conserve the strangeness, the Ω^- decay occurs trough weak interaction. The favorite channel for the weak decay of the s quark is:

$$\Omega^- = (ss)s \to (ss)uW^- \to \big[(ss)u\big][d\overline{u}] = \Xi^0\pi^-$$

the (ss) pair in brackets represents spectator quarks. As the path length of the Ξ^0 is similar to that of the Ω^-, its lifetime is also of the order of 10^{-10} s.

The Ξ^0 decays into three particles: two of them are γ-rays (denoted with γ_1, γ_2 in the picture), and each of them produces an e^+e^- pair. The γ_1, γ_2 are likely produced by a $\pi^0 \to \gamma\gamma$ decay. The π^0 lifetime is $\sim 10^{-16}$ s and its path length of $\sim 10^{-6}$ cm is not visible in the picture. The particle almost collinear with the Ξ^0 must be a hadron. The Ξ^0 decay is a $\Delta S = 1$ weak interaction process:

$$\Xi^0 = (su)s \to (su)uW^- \to (su)u[d\overline{u}] = \big[(su)d\big][u\overline{u}] = \Lambda^0\pi^0$$

The Λ^0 has a path length (and a corresponding lifetime) 2–3 times larger than that of the Ω^- and of the Ξ^0.

Finally, the Λ^0 decays into two charged particles: a π^- and a proton. Its $\Delta S = 1$ decay channel is:

$$\Lambda^0 = (ud)s \to (ud)uW^- \to (ud)u[d\overline{u}] = \big[u(ud)\big][d\overline{u}] = p\pi^-$$

The proton is stable; the charged pions have lifetime sufficient to escape the region of the picture.

Problem 8.12

(a) The decay $\Lambda^0 \to pe^-\overline{\nu}_e$ is a semileptonic weak decay of the Λ baryon. Because of the $s \to u$ flavor transition, this decay is Cabibbo suppressed and the matrix element is proportional to $g_{e\nu}g_{us} = G_s = G_F \sin\theta_C$ (see Eq. (8.48c)).

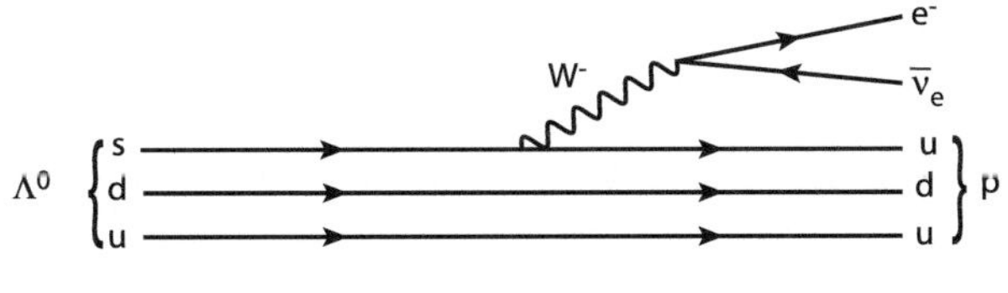

(b) The $\Xi^- \to \Lambda^0\pi^-$ is a nonleptonic weak decay. In terms of quarks, it corresponds to the same $s \to u$ transition as in the case (a). The W^- here decays into a $\overline{u}d$ quark pair. The matrix element is proportional to $g_{ud}g_{us} = G_F \sin\theta_c \cos\theta_c$

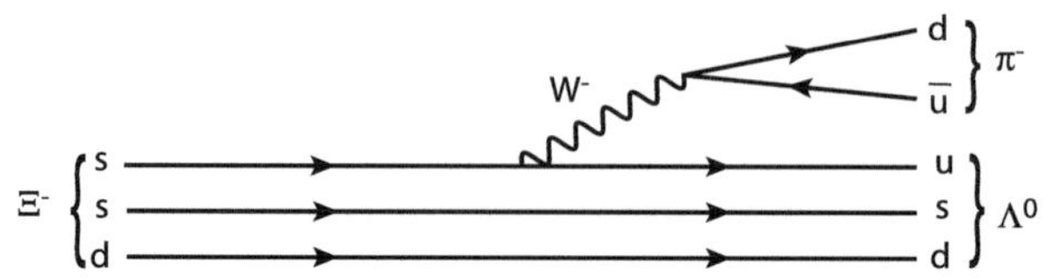

(c) The $\nu_\mu p \to \mu^- \Delta^{++}$ reaction is a weak interaction process with the exchange of a W^- boson. The coupling at the quark vertex is g_{ud} and at the lepton vertex is $g_{\mu\nu} = g_{e\nu}$ for the lepton universality. The matrix element is proportional to $g_{ud} g_{e\nu} = G_F \cos\theta_c$

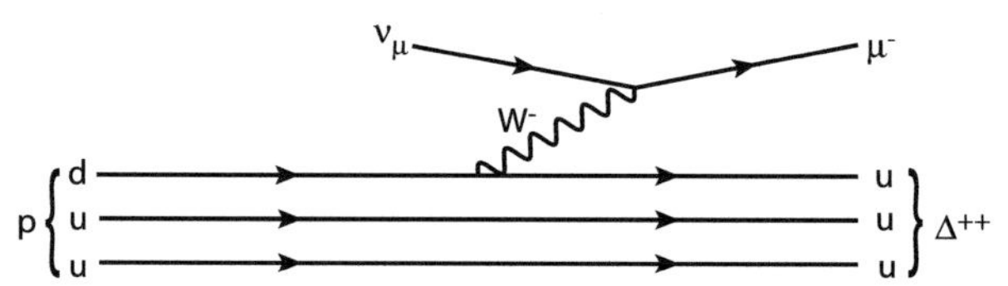

(d) The $D^0 \to K^- \mu^+ \nu_\mu$ is a weak decay with a $c \to s$ flavor transition. The matrix element is proportional to $g_{e\nu} g_{cs} = G_F \cos\theta_C$ (see Eq. (8.58a)). The c quark prefers to decay (by weak interaction) in the s quark, see Fig. 8.20.

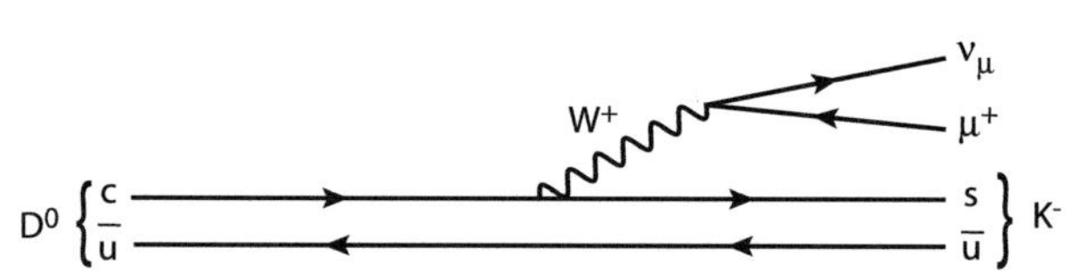

(e) The $D^0 \to K^+ \mu^- \overline{\nu}_\mu$ decay is strongly suppressed. In fact (referring to the above $D^0 \to K^- \mu^+ \nu_\mu$ diagram) as quark $\to$ antiquark transitions are forbidden, it requires a *second-order* weak interaction transition $\overline{u} \to \overline{s}$ and $c \to u$. Second-order means that two W bosons must be exchanged in the Feynman diagram.

Problem 8.14 In the following table, from left to right are reported: the W^+ decay channel; the square of the CKM coupling times the color factor; the relative decay fraction. The decay channels with a t quark are forbidden, as $m_W < m_t$.

Decay	$N_c \times V_{ij}^2$	$BR(\%)$
$W^+ \to u\bar{d}$	$N_c \times V_{ud}^2 = 3 \cdot 0.974^2 = 2.846$	31.5%
$W^+ \to c\bar{d}$	$N_c \times V_{cd}^2 = 3 \cdot 0.225^2 = 0.152$	1.8%
$W^+ \to u\bar{s}$	$N_c \times V_{us}^2 = 3 \cdot 0.225^2 = 0.152$	1.8%
$W^+ \to c\bar{s}$	$N_c \times V_{cs}^2 = 3 \cdot 0.973^2 = 2.840$	31.5%
$W^+ \to u\bar{b}$	$N_c \times V_{ub}^2 = 3 \cdot 0.0035^2 = 3.6 \cdot 10^{-5}$	$4 \cdot 10^{-6}$
$W^+ \to c\bar{b}$	$N_c \times V_{cb}^2 = 3 \cdot 0.041^2 = 0.048$	0.1%
$W^+ \to e^+\nu_e$	$V^2 = 1$	11.1%
$W^+ \to \mu^+\nu_\mu$	$V^2 = 1$	11.1%
$W^+ \to \tau^+\nu_\tau$	$V^2 = 1$	11.1%
Sum	9.038	100%

It is easy to verify that the W decays into hadrons represent 67% of the cases, while the leptonic decay amounts to 33%.

Problem 8.15 A Lorentz transformation between two inertial frames along the x axis with relative velocity $\beta = v/c$ can be expressed as:

$$x'_\mu = \sum_\nu a_{\mu\nu}(\beta)x_\nu; \quad a_{\mu\nu} = \begin{pmatrix} (1-\beta^2)^{-1/2} & -\beta(1-\beta^2)^{-1/2} & 0 & 0 \\ \beta(1-\beta^2)^{-1/2} & (1-\beta^2)^{-1/2} & 0 & 0 \\ 0 & 0 & 1 & 0 \\ 0 & 0 & 0 & 1 \end{pmatrix}.$$

In Appendix 4, we show that the Dirac equation is a relativistic invariant if its solutions are represented by functions that satisfy the relation:

$$\psi' = S\psi; \quad \text{with } S^{-1}\gamma^\mu S = \sum_\nu a_{\mu\nu}\gamma^\nu$$

where S is an operator that acts only on the spinor part of the wave function ψ.

(a) Regarding the relativistic invariance of the scalar:

$$\overline{\psi}\psi \xrightarrow{Lorentz} \overline{\psi}'\psi' = (\overline{\psi}S^{-1})(S\psi) = \overline{\psi}(S^{-1}S)\psi = \overline{\psi}\psi.$$

which shows that the quantity $\overline{\psi}\psi$ does not depend on the reference system, as required for a scalar quantity.

(b) Regarding the relativistic invariance of the four-vector:

$$\overline{\psi}\gamma^\mu\psi \xrightarrow{Lorentz} \overline{\psi}'\gamma^\mu\psi' = (\overline{\psi}S^{-1})\gamma^\mu(S\psi) = \overline{\psi}(S^{-1}\gamma^\mu S)\psi$$
$$= \sum_\nu a_{\mu\nu}\overline{\psi}\gamma^\mu\psi\overline{\psi}\psi.$$

The last equality shows that the quantity $\overline{\psi}\gamma^\mu\psi$ is correctly transformed as a space-time four-vector.

Problem 8.16 Neglecting the electron mass, one has: $E_e = p_e c$; therefore, one can write:

$$\int_0^{E_0/c} p_e^2(E_0 - p_e c)^2 dp_e$$

$$= \int_0^{E_0/c} p_e^2\left(E_0^2 + p_e^2c^2 - 2E_0 p_e c\right)dp_e$$

$$= \left[\frac{p_e^3 E_0^2}{3} + \frac{p_e^5 c^2}{5} - \frac{2E_0 p_e^4 c}{4}\right]_0^{E_0/c} = \left[\frac{20+12-30}{60}\frac{E_0^5}{c^3}\right] = \frac{E_0^5}{30c^3}.$$

Problem 8.17

(a) The branching ratio for the leptonic decays are almost equal because $m_\tau \gg m_\mu, m_e$ and the energies available in the final state are almost the same. Due to the lepton universality, the matrix elements for the $e^-\overline{\nu}_e$ and $\mu^-\overline{\nu}_\mu$ are the same.

(b) The decay into hadrons takes place through a $q\overline{q}$ pair. Only combinations of quarks u, d, s are allowed, as the c, b, t hadrons have masses larger than that of the τ. The transition probability for the $\tau^- \to d\overline{u} + \nu_\tau$ process is proportional to $G_F^2 \cos^2\theta_c \simeq 0.95 G_F^2$, where $\theta_C \simeq 0.23$ is the Cabibbo angle. The transition probability for $\tau^- \to s\overline{u} + \nu_\tau$ is proportional to $G_F^2 \sin^2\theta_c \simeq 0.05 G_F^2$. The difference between the particle masses in the final state is small compared to the τ mass, and the phase space factor is almost the same for the decay into quarks or leptons.

Each quark flavor, however, has three degrees of freedom (N_c) due to the *color* quantum number. The expected ratio between the decay probabilities into hadrons or muons is:

$$\frac{\Gamma(\tau^- \to hadrons + \nu_\tau)}{\Gamma(\tau^- \to \mu^-\overline{\nu}_\mu\nu_\tau)} \simeq N_c \times \frac{\Gamma(\tau^- \to d\overline{u}\nu_\tau)}{\Gamma(\tau^- \to \mu^-\overline{\nu}_\mu\nu_\tau)} = 3 \times \frac{G_F^2 \cos\theta_c^2}{G_F^2} \simeq 2.85$$

The experimental fraction due to all hadronic decays is obtained by adding the different final state channels: a hadron and a neutrino; a hadron plus at least one neutral particle and a neutrino:

$$\Gamma(\tau^- \to hadrons + \nu_\tau) = \Gamma(h^-\nu_\tau) + \Gamma(h^- \geq 1\ neutral\ \nu_\tau)$$
$$= 0.12 + 0.37 = 0.49$$

so that, one has:

$$\frac{\Gamma(\tau^- \to hadrons\ \nu_\tau)}{\Gamma(\tau^- \to \mu^-\overline{\nu}_\mu\nu_\tau)} = \frac{0.49}{0.174} = 2.8$$

in good agreement with the expected value.

Fig. 8.12 Example of a Feynman diagram with a $\Delta S = 1$ neutral current: $K^0 \to \mu^+\mu^-$

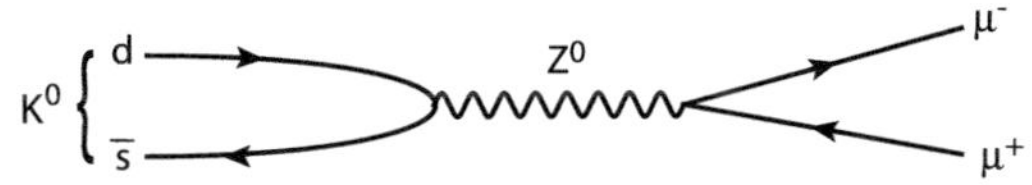

(c) Experimentally the ratio $\frac{\Gamma(\tau^- \to K^- \nu_\tau)}{\Gamma(\tau^- \to \pi^- \nu_\tau)} = \frac{0.696}{10.91} = 0.064$.

The decay reported in the denominator is a decay $\tau^- \to d\overline{u} + \nu_\tau \propto G_F^2 \cos^2\theta_c$. The decay reported in the numerator instead is of the type $\tau^- \to s\overline{u} + \nu_\tau \propto G_F^2 \sin^2\theta_c$. The expected ratio is therefore $\simeq \frac{0.05}{0.95} = 0.053$.

Problem 8.18 A Feynman diagram for the possible $K^0 \to \mu^+\mu^-$ decay would necessarily imply the transformation of a quark pair into leptons via a neutral current weak interaction (see Fig. 8.12). This would correspond to a $\Delta S = 1$ transition. The non-observation of such processes (with a branching ratio $BR < 10^{-7}$) has led to the hypothesis that $\Delta S = 1$ neutral currents are not allowed. From the theoretical point of view, this required the introduction of a new quark (the quark c, see Sect. 8.14.2).

Problem 8.19 The lifetimes for the three particles are (from PDG [P10]): $\tau(D^+) = 1040 \times 10^{-15}$ s, $\tau(D^0) = 410 \times 10^{-15}$ s and $\tau(D_s) = 500 \times 10^{-15}$ s. As the strong interaction conserves the flavor, the allowed decay modes are due to the weak interaction (as evident from the long lifetimes). In the spectator quark model of weak decays, it is assumed that the non-decaying, or *spectator*, quarks do not interact with products of the virtual decay of the other. The three particles are charmed mesons: the c quark can decay:

$$c \to s \propto \cos^2\theta_c; \qquad c \to d \propto \sin^2\theta_c$$

The amplitude probability for the decay of the c quark in the D^0, D^+, D_s^+ is the same for the three mesons.

In the non-strange D^0, D^+ mesons, the small mass difference between the u or d quarks allows the decay:

$$d \to u \quad \text{and} \quad u \to d \propto \cos^2\theta_c$$

but the phase-space factor is small.

In the strange D_s^+ meson, the only allowed decay is:

$$s \to u \propto \sin^2\theta_c$$

but, due to the large mass difference between the two quarks, the phase-space factor is large. As the decay probability is the product of the amplitude probability and the phase-space factor, there is a compensation effect for the u, d and s decays of the second quark in the D mesons.

References

[8B64] Barnes, V.E., et al.: Observation of a hyperon with strangeness number three. Phys. Rev. Lett. **12**, 204 (1964)

[8C97] Reines, F., Cowan, C., Jr.: Reines-Cowan experiments: detecting the poltergeist. Los Alamos Science n. 25 (1997). http://la-science.lanl.gov/lascience25.shtml

[8T10] Tavernier, S.: Experimental Techniques in Nuclear and Particle Physics. Springer, Berlin (2010). ISBN: 978-3642008283

Chapter 9
Discoveries in Electron–Positron Collisions

Problems

9.1. **The strong coupling constant**. Using the data shown in Fig. 9.4 and the potential given in Eq. (9.16), estimate the value of the strong coupling constant α_s at the energy of charmonium formation. Assume that $m_c c^2 = 1550$ MeV.
[See solutions]

9.2. **Event rate in e^+e^- collider**. In a small electron-positron collider of $R = 10$ m radius, each beam has a current intensity of $I = 10$ mA and a transverse area of $S = 0.1$ cm^2.
(a) Assuming that each of the two e^- and e^+ beams is contained in one single bunch and that the beams collide head-on twice per revolution, calculate the collider luminosity in cm^{-2} s^{-1}.
(b) The cross-section for the reaction $e^+e^- \to \pi^+\pi^-\pi^0$ at the peak of the ω resonance is $\sigma = 1.5$ µb. Calculate the number of observed events per hour for this process.
[See solutions]

9.3. **Beam attenuation in the beam pipe**. The vacuum in the pipe of an electron accelerator has a pressure of $p = 3 \cdot 10^{-4}$ tor. The electron beam corresponds to an average current of $I = 60$ mA. If a one meter long ($l = 1$ m) section is considered, assuming that the residual gas inside the vacuum pipe is made of hydrogen atoms and knowing that the $e^-p \to e^-p$ cross-section is $\sigma = 1$ µb, calculate the number of beam-gas collision events expected in one second.
[Suggestion: see Problem 3.16]

9.4. **Design LEP luminosity**. Determine the theoretical luminosity (see Sect. 3.3) of the LEP accelerator using the following parameters: $n_p = n_a = 4$, number of circulating particle/antiparticle bunches; $f = 11240$ Hz, revolution frequency;

S. Braibant et al., *Particles and Fundamental Interactions: Supplements, Problems and Solutions*, Undergraduate Lecture Notes in Physics,
DOI 10.1007/978-94-007-4135-5_9, © Springer Science+Business Media Dordrecht 2012

$I = 1$ mA, intensity of the circulating current per beam; $\sigma_h = 300$ μm, beam width; $\sigma_v = 8$ μm beam height.
[See solutions]

9.5. **LEP luminosity from the forward detector**. Determine the LEP luminosity at $\sqrt{s} = 91$ GeV using the experimental information of a forward detector able to measure elastic $e^+e^- \to e^+e^-$ collisions. The forward detector covers the solid angle: 40 mrad $< \theta <$ 150 mrad, $0 < \varphi < 2\pi$, and the measured $e^+e^- \to e^+e^-$ event rate is $R = 0.20\ \mathrm{s}^{-1}$.
[A: $\frac{(8\pi\alpha^2\hbar^2c^2)}{E^2}\left(\frac{1}{1-\cos\theta_{min}} - \frac{1}{1-\cos\theta_{max}}\right) = 3\cdot 10^{30}\ \mathrm{cm}^{-2}\,\mathrm{s}^{-1}$]

9.6. **Radiative return to the $\mathbf{Z^0}$**. In the case of an e^+e^- collider, the effective c.m. energy $\sqrt{s}'$ is smaller than the initial one, $\sqrt{s}$, if a photon is emitted from the primary positron or electron. Assuming that the photon energy is E_γ, calculate s'. How does an event with two hadronic jets in the final state look like in this case?
[See solutions]

9.7. ***J/ψ* resonance**. The intrinsic width of the J/ψ resonance is smaller than the experimental resolution (which was about 2 MeV in the first experiments). It can be indirectly obtained from quantities that do not depend on the resolution. Consider the $e^+e^- \to J/\psi \to e^+e^-$ reaction. The measured cross-section integrated over the resonance is $\int_{resonance} \sigma_{e^+e^-} d\sqrt{s} \simeq 790$ nb MeV. Derive the value of the J/ψ intrinsic width ($\Gamma_{J/\psi}$) knowing that the mass of the resonance is $M_{J/\psi} \simeq 3097$ MeV and that the branching ratio for the decay into e^+e^- is $BR(J/\psi \to e^+e^-) \simeq 0.06$.
[See solutions]

Supplement 9.1: Electronic Logic and Trigger

Electronic Logic Figure 9.1 shows the definitions, graphical symbols and the truth tables of the simplest electronic circuits. The corresponding *Boolean logic expressions* are also given. Remember the following logical circuits (or *gates*):

NOT: provides an output signal which is true if the input is false.
AND: provides an output signal when a signal is simultaneously present on both input channels.
OR: provides an output signal when a signal is present on at least one of the two input channels.
XOR: provides an output signal if A is not equal to B.
NAND: provides an output signal if A and B are not both true.
NOR: provides an output signal if neither A nor B is true.

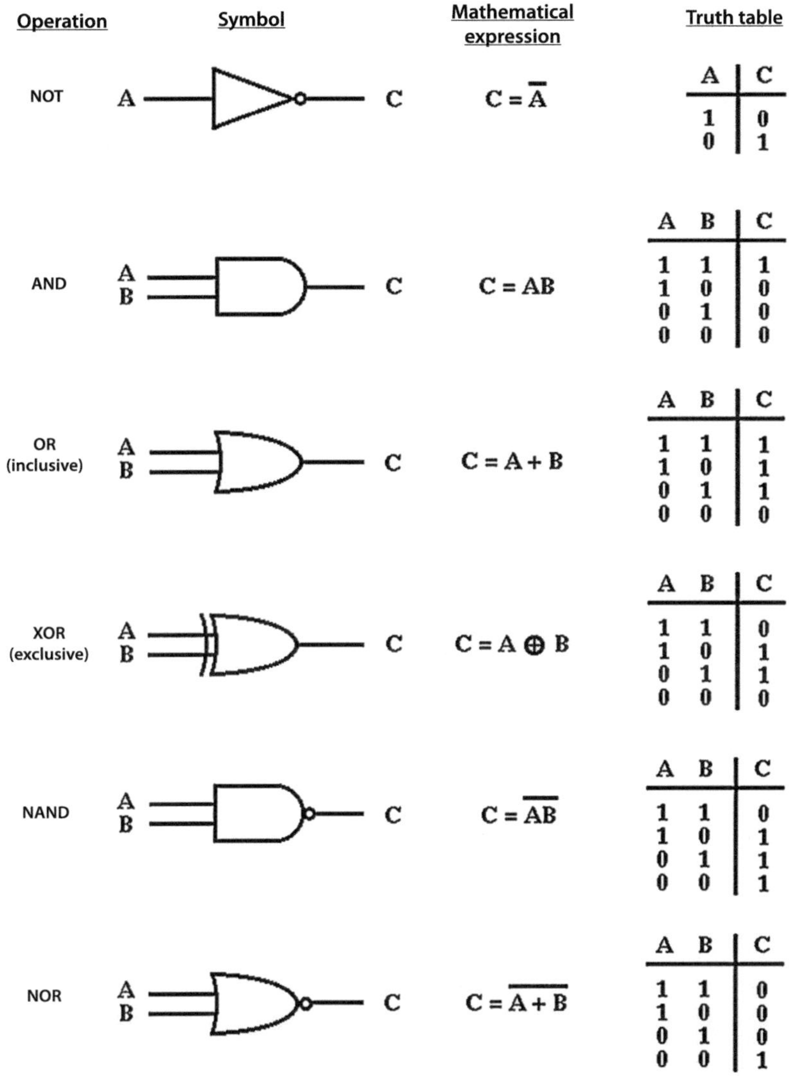

Fig. 9.1 Definitions, symbols, Boolean expressions and truth tables for some examples of logical circuits ("gates")

Trigger In any experiment, particular reactions must be disentangled from background or other reactions. To do this, specific criteria that identify the reaction are imposed. The requested electronic logic provides a *trigger* that activates the recording of the event.

Trigger Example Figure 9.2a shows a simple layout of an pion-nucleon elastic scattering experiment as performed in the 1960s with only 4 scintillators. The trigger

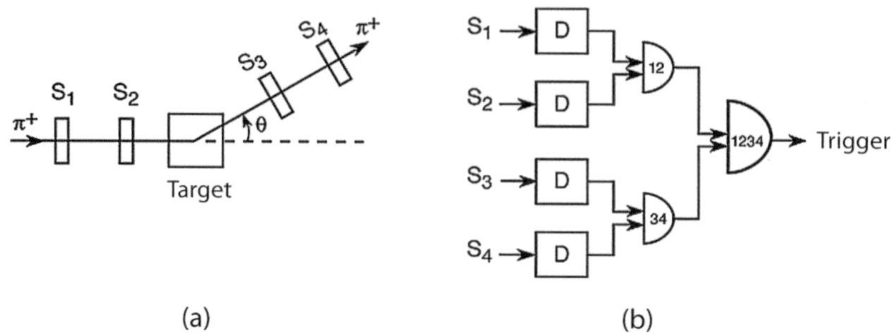

Fig. 9.2 (**a**) Layout of a simple elastic scattering experiment $\pi^+ p \to \pi^+ p$ using 4 scintillator counters S_1, S_2, S_3, S_4. (**b**) Trigger layout: the incident π^+ is defined by the coincidence $(S_1 S_2)$; the π^+ scattered at the angle θ is defined by the coincidence $(S_3 S_4)$. The final trigger is defined by the coincidence $(S_1 S_2 S_3 S_4)$

is defined as the coincidence in time $(S_1 S_2 S_3 S_4)$ that identifies the incident pion and the pion scattered at an angle θ. When there is a coincidence in $(S_1 S_2)$ and $(S_3 S_4)$, one has a trigger, and the recording of the pulse heights in the S_1, S_2, S_3, S_4 counters is immediately activated. The concordance of $(S_1 S_2)$ and NOT $(S_3 S_4) = (\overline{S}_3 \overline{S}_4)$ would represent an incident pion without scattering at that particular angle. The coincidence of $(S_1 S_2)$ and $(S_3 S_4)$ represents the searched π^+ scattering at the angle θ. A $(\overline{S}_1 \overline{S}_2)$ and $(S_3 S_4)$ would represent a *spurious* signal.

In more complex experiments, where one wants to measure several reactions at the same time, the final trigger is an OR of several different triggers. In the case of the measurement of rare events, you can have triggers of 1st and of 2nd level: the 1st level trigger performs a loose selection; the 2nd level trigger uses more accurate criteria. For the LHC, the trigger logic is one of the most important and delicate aspects of the experiment (see Supplement 10.1).

Electronic Standards Let us consider the following standards:

NIM (Nuclear Instrument Module) is a modular standard, in which electronic circuits (discriminators, coincidence units, amplifiers, etc.) are modules with predefined mechanical and electrical specifications; these modules are inserted in standard "slots" (bins) of a "NIM" crate that contains all the necessary power supplies. In this standard, it is simple to assemble (and replace if necessary) the modules needed to perform an experiment. This standard logic is a voltage (−0.6 V) or current based logic, with a rise time of about 1 ns and a pulse width $\geq$10 ns into impedances of 50 Ω.

TTL (Transistor-Transistor Logic) and **ECL (Emitter-Coupled Logic)** are two other electronic standards whose modules can be inserted in the NIM standard.

CAMAC was originally designed as the standard to use in conjunction with the NIM standard to connect with computers. All current experiments in particle physics use computers for their *data acquisition system*. Over time, the acquisition system became more complex. It not only records data, but it also carries out functions of monitoring, calibration, on-line display and analysis of data samples, etc. The NIM is not suitable to perform these functions, while the CAMAC can execute most of them although it is insufficient when the amount of data is too large. In this case, the **FASTBUS** standard is preferred. The CAMAC (as FASTBUS) is a modular system made of fundamental elements such a CAMAC crate and several plug-in modules that can be inserted into specific "slots". The backplane of the CAMAC crate contains a set of connections forming the *dataway* or the *backplane bus* that connects all the plug-units between themselves, so that data can be exchanged between the various modules and sent to the computer, under the supervision of some *controller modules.*

The broad use of microcomputers at all levels of the electronics, with temporary memories, logic triggering systems, etc., has required new connection systems, new data transfer methods, etc. The most used standard is the **VME** standard that is used in conjunction with CAMAC and FASTBUS.

Computers The use of computers is in constant development, with the general trend towards distributed systems of computers connected in networks and in networks of networks. The data acquisition systems tend to use more and more the *World Wide Web* (WWW) methodology with reference databases, transfer of multimedia information and on-line monitoring. These methods are also used in maintaining communications within large collaborations, but also to the outside world.

Solutions

Problem 9.1 A positron and an electron can form a bound state (the so-called *positronium*), whose energy levels, described in Fig. 9.4a, are similar to those of the hydrogen atom. The positronium can be completely described by quantum mechanics, since the potential $V_{em} = -\alpha_{EM}/r$, where r is the distance between the positron and the electron, is known. The energy levels can be obtained by solving the Schrödinger equation, or with the quantization rules of Bohr's atom. One finds that (n is the main quantum number):

$$E_n = -\frac{\alpha_{EM}^2 m_e c^2}{4n^2} \tag{9.1}$$

Note that there is a factor of 2 difference compared to the energy levels of hydrogen atoms. This is due to the fact that the masses m_e of the electron and positron are identical and the *reduced mass* $\mu = m_e/2$ must be introduced.

For the *charmonium* (bound state $c\bar{c}$), the interaction between the quark and the antiquark can be described by the potential given in Eq. (9.16), neglecting the second term which dominates only for large values of r (that is certainly not the case for bound states). (For a full discussion on the spectrum of hadrons containing heavy c and b quarks, see [9D95].) This implies that the solution of the Schrödinger equation of the hydrogen atom shall yield the same eigenvalues as for the charmonium, with the substitution $\alpha_{EM} \rightarrow (4/3)\alpha_S$. This allows a rough estimate of α_S using the differences in energy levels shown in Fig. 9.15b.

Consider the states with $n = 1$, $n = 2$. In the figure, they are indicated as 1^1S_0, 2^1S_0. The energy difference between the two levels is about $\Delta E = 500 - (-100)$ MeV $= 600$ MeV. Using the relation for the energy levels E_n (replacing $m_e \rightarrow m_c$ and $\alpha \rightarrow 4/3\alpha_s$), one obtains:

$$\Delta E = E_2 - E_1 = \left(\frac{4\alpha_s}{3}\right)^2 \frac{m_c c^2}{4}\left(1 - \frac{1}{4}\right).$$

Using $m_c c^2 = 1550$ MeV and the measured difference $\Delta E \simeq 600$ MeV, one finds:

$$\alpha_S^2 = \frac{3\Delta E}{m_c c^2} = \frac{1800}{1500} \simeq 1.2$$

This is a rough estimate, showing that the value of α_S is very large compared to the electromagnetic constant [P95].

Quantum Chromodynamics (QCD), the gauge field theory that describes the strong interaction of colored quarks and gluons, is the SU(3) component of the $SU(3)_C \times SU(2)_L \times U(1)_Y$ group theory of the Standard Model of Particle Physics. The QCD Lagrangian can be formally obtained in this framework (this topic is outside the formalism level of this book. Refer for instance to Sect. 9.1 of [P10]). In this QCD Lagrangian, there is only one free parameter, the strong coupling constant α_S, if the quark masses are fixed. The coupling constant in itself is not a physical observable, but rather a quantity defined in the context of perturbation theory (at high transferred momentum) or lattice QCD (for low transferred momentum), which gives predictions for experimentally measurable observables.

An example of QCD lattice calculation is the value of α_S from the charmonium energy levels is reported in [9A92]. See also Problem 11.5 and [9B96].

Problem 9.2

(a) A current of $I = 10$ mA corresponds to the fact that a number N_e of accelerated charged particles move (almost at the light speed c). Taking into account that the revolution period inside the accelerator is $T = (2\pi R/c) = 2.1 \times 10^{-7}$ s and that the electric charge of electrons and positrons (in absolute value) is $e = 1.6 \times 10^{-19}$ C, one has:

$$I = (N_e \cdot e/T) \text{ C/s}.$$

The corresponding number of e^+/e^- in each beam is then:

$$N_e = \left[2\pi R \cdot I/(c \cdot e)\right] = 1.3 \times 10^{10}.$$

The machine luminosity (considering $n = 2$ collisions per revolution, occurring with frequency $f = 1/T$) is:

$$\mathcal{L} = \left(\frac{N_{e^+} N_{e^-} n f}{S}\right) = \frac{2N_e^2}{TS} = \frac{2 \cdot (1.3 \times 10^{10})^2}{2.1 \times 10^{-7} \cdot 0.1} = 1.6 \times 10^{28} \text{ cm}^{-2}\,\text{s}^{-1}.$$

(b) The event rate R is obtained by multiplying the machine luminosity by the process cross-section (1 b=10^{-24} cm^2):

$$R = \mathcal{L}\sigma = \left(1.6 \times 10^{28} \cdot 1.5 \times 10^{-30}\right) = 0.024\ \text{s}^{-1}.$$

Taking into account the number of seconds in one hour, there are 87 events/hour.

Problem 9.4 The luminosity $\mathcal{L}$ can be computed from the parameters of the accelerator, using Eq. (3.11b):

$$\mathcal{L} = \frac{f n_p n_a N_p N_a G}{4\pi \sigma_h \sigma_v}.$$

Here, $n_a = n_p = 4$, $f = 11240$ Hz (which corresponds to the number of revolutions completed per second by particles in the 27 km long LEP vacuum tube). The factor G is assumed $= 1$. The quantity r^2 in Eq. (3.11b) used for a beam of circular transverse section is replaced by $\sigma_h \sigma_v$ for the beam ellipsoidal section. The current $I = 1$ mA corresponds to a number N_e of particles with electric charge $e = 1.6 \times 10^{-19}$ C (see Problem 9.2) equal to $N_e = I/(e \cdot f) = 0.6 \times 10^{12}$.

Inserting the numerical values $[4\pi\sigma_h\sigma_v = 4\pi(300 \times 10^{-4})(8 \times 10^{-4}) = 3 \times 10^{-4}\ \text{cm}^2]$, one finds:

$$\mathcal{L} = \frac{11240\ [\text{s}^{-1}] \cdot (0.6 \times 10^{12})^2 \cdot 4^2 \cdot 1}{3 \times 10^{-4}\ [\text{cm}^2]} = 2 \times 10^{32}\ \text{cm}^{-2}\,\text{s}^{-1}.$$

Problem 9.6 In a collider, the mass-energy relativistic invariant is:

$$s = \left[(E, \mathbf{p}) + (E, -\mathbf{p})\right]^2 = (2E)^2.$$

If, before the collision, the electron or the positron radiates a real photon with four-momentum $(h\nu, \mathbf{h}\nu)$:

$$\begin{aligned} s' &= \left[(E - h\nu, \mathbf{p} - \mathbf{h\nu}) + (E, -\mathbf{p})\right]^2 \\ &= (2E - h\nu; -\mathbf{h\nu})^2 = (2E)^2 - 2h\nu(2E). \end{aligned}$$

Then, one has:

$$s' = s - 2h\nu\sqrt{s} = s - 2E_\gamma\sqrt{s}.$$

A radiative event with two hadronic jets in the final state is due to the Z^0 decay in a quark-antiquark pair: $e^+e^- \to e^+e^-\gamma \to Z^0\gamma \to q\overline{q}\gamma$. As in the radiative event shown in Fig. 9.14 (where a $\mu^+\mu^-$ pair is produced in the final state), the vector sum of the momenta of the three particles observed in the final state (i.e., the two hadronic jets and the radiated photon) has to be equal to zero. As a consequence, the two hadronic jets are not collinear (not *back-to-back*).

Problem 9.7 The dependence of the J/ψ production cross-section and its decay into a given channel as a function of energy is given by the Breit-Wigner formula given in Eq. (9.11):

$$\sigma(\sqrt{s})_{(e^+e^-\to J/\psi\to e^+e^-)} = \frac{\pi\lambda\!\!\!^{-2}(2J+1)\Gamma_{ee}\Gamma_{ee}}{(2s_1+1)(2s_2+1)[(\sqrt{s}-E_R)^2+\Gamma^2/4]}$$
$$= \frac{3\pi\lambda\!\!\!^{-2}\Gamma_{ee}^2}{4[(\sqrt{s}-E_R)^2+\Gamma^2/4]} \tag{9.2}$$

with $J=1, s_1=s_2=1/2$. Compared to Eq. (9.11), we substituted $\Gamma_h \to \Gamma_{ee}$ to take into account that the problem gives the BR for the J/ψ decay into an e^+e^- pair: $\Gamma_{ee}=0.06\Gamma$. This corresponds to the fact that only 6% of the observed J/ψ events decay into an e^+e^- pair. Then, Eq. (9.2) becomes:

$$\sigma(\sqrt{s}) = \frac{3\pi\lambda\!\!\!^{-2}\Gamma_{ee}^2}{4}\frac{1}{[(\sqrt{s}-E_R)^2+\Gamma^2/4]}. \tag{9.3}$$

The total width Γ of the resonance can be calculated with the above formula, providing that $\int\sigma(\sqrt{s})d\sqrt{s} = 790$ nb MeV. By integrating the energy-dependent part of (9.3), one obtains:

$$\int_{-\infty}^{\infty}\frac{1}{[(\sqrt{s}-E_R)^2+(\Gamma/2)^2]}d\sqrt{s}$$
$$= \int_{-\infty}^{\infty}\frac{1}{(\Gamma/2)[(\frac{\sqrt{s}-E_R}{\Gamma/2})^2+1]}d\left(\frac{\sqrt{s}-E_R}{\Gamma/2}\right)$$
$$= \int_{-\infty}^{\infty}\frac{1}{(\Gamma/2)(x^2+1)}dx = \frac{2}{\Gamma}\int_{-\infty}^{\infty}\frac{1}{(x^2+1)}dx = \frac{2}{\Gamma}[\mathrm{arctg}\,x]_{-\infty}^{\infty} = \frac{2\pi}{\Gamma}.$$

From (9.3), one has:

$$\int\sigma(\sqrt{s})d\sqrt{s} = \frac{3\pi^2\lambda\!\!\!^{-2}(0.06)^2\Gamma^2}{2\Gamma} = 790 \text{ nb MeV}.$$

The De Broglie wavelength can be computed remembering that the momentum at the J/ψ production is $p=E_R/2=1548$ MeV/c, corresponding to a De Broglie wavelength of $\lambda\!\!\!^- = \hbar/p = 6.6\times10^{-22}$ (MeV s)/1548 (MeV/c) $=1.27\times10^{-14}$ cm. Inserting the numerical values, one finds:

$$\Gamma = \frac{2\cdot 790 \text{ nb MeV}}{3\pi^2\lambda\!\!\!^{-2}(0.06)^2} = 0.093 \text{ MeV}$$

The J/ψ intrinsic width of ~93 keV is much lower than the measured width due to the finite resolution of experimental apparata.

References

[9A92] El-Khadra, A.X., et al.: Determination of the strong coupling constant from the charmonium spectrum. Phys. Rev. Lett. **69**, 729–732 (1992)

[9D95] Davies, C.T.H.: The heavy hadron spectrum. Lectures at Schladming Winter School (1997). arXiv:hep-ph/9710394v1

[9B96] Burrows, P.N., et al.: Prospects for the precision measurement of α_S. arXiv:hep-ex/9612012v2

Chapter 10
High Energy Interactions and the Dynamic Quark Model

Problems

10.1. **De Broglie wavelength**. Using the de Broglie relation, determine the momentum p of a probe particle needed to solve the structure of:
(a) an iron nucleus ($A = 56$);
(b) a nucleon.
(c) Estimate the particle energy needed to probe the size r_q of quarks, knowing that the experimental upper limit is $r_q < 10^{-16}$ cm.
[A: (a) 270 MeV/c; (b) 1 GeV/c ; (c) >1.2 TeV/c]

10.2. **Electron inelastic scattering**. An electron with a $E = 20$ GeV kinetic energy collides inelastically on a proton at rest. The electron is scattered at an angle $\theta = 5°$ with respect to its original direction and with an energy $E' = 12$ GeV. Calculate the effective mass of the final hadronic system.
[See solutions]

10.3. **Structure function**. The momentum distribution of the u-type quark in the proton can be parameterized by the formula $F_u(x) \simeq xu(x) = a(1-x)^3$. Determine the constant a with the assumption that the u quarks carry 33% of the proton momentum.
[See solutions]

10.4. **Quark distribution**. The distributions of u quarks in the proton and of $\overline{d}$ antiquark in the antiproton can be assumed to be represented by the functions: $F_u(x) = xu(x) = a_1(1-x)^3$, $F_{\overline{d}}(x) = x\overline{d}(x) = a_2(1-x)^3$, where x is the Bjorken variable, i.e., the fraction of the nucleon momentum carried by quarks. Assuming that the quarks contribute to half of the nucleon momentum, calculate the constant a_1 and a_2.
[A: $a_1 = 4/3$, $a_2 = 2/3$].

S. Braibant et al., *Particles and Fundamental Interactions: Supplements, Problems and Solutions*, Undergraduate Lecture Notes in Physics,
DOI 10.1007/978-94-007-4135-5_10,

10.5. **Gluon structure function**. It is believed that the structure function describing the distribution of the gluon momentum inside the nucleons, $g(x)$, strongly increases with decreasing x. Estimate the number of gluons that would be possible to resolve with deep inelastic $e + p \to e + X$ collisions at $Q^2 = 10^4$ GeV^2 at low x values (in the intervals $0.0001 \div 0.001$, $0.001 \div 0.01$, $0.01 \div 0.1$). Assume that at these Q^2 values the distribution function of the gluons is $g(x) = 0.36x^{-0.5}$.
[See solutions]

10.6. **$W^{\pm}$, Z^0 production at the S$p\overline{p}$S CERN collider**. The UA1 and UA2 experiments at the CERN S$p\overline{p}$S collider led to the discovery of the $W^{\pm}$, Z^0 vector bosons of the weak interaction. At the CERN S$p\overline{p}$S collider, protons and antiprotons were made to collide with a total c.m. energy of 540 GeV (later, 630 GeV).

Discuss in terms of quark interactions at which energy the $W^{\pm}$, Z^0 production is obtained.
[See solutions]

10.7. **Neutrino/antineutrino cross-section ratio**. The ratio $R = \sigma_{\overline{\nu}}/\sigma_{\nu}$ between the cross-sections of neutrinos and antineutrinos colliding on an isoscalar target (i.e., a target with an equal number of protons and neutrons) is $R \simeq 0.5$ (see Fig. 10.12). This experimental result cannot be easily accounted for in the framework of the static quark model of hadrons, since it requires that a small fraction of the nucleon momentum be carried by antiquarks. In the dynamic quark model, many sea quark-antiquark pairs are present inside the nucleon (see Fig. 7.21).

Using the measured value of R, and taking into account the coupling given in Eq. (10.57) of neutrinos and antineutrinos with quarks and antiquarks, determine the ratio between the fraction of the momentum carried by quarks and antiquarks in the nucleon.
[See solutions]

10.8. **Quark and antiquark content of ordinary matter**. Using the integral of the quark and antiquark distribution functions given in Eq. (10.67), determine the ratio between the fraction of the momentum carried by quarks and antiquarks in isoscalar nuclei. Compare with the result obtained in the previous problem.
[See solutions]

10.9. **Neutrino beams-1**. Muon neutrino beams are extensively used in many experimental situations, for instance in deep inelastic scattering as well as in long and short baseline neutrino oscillation experiments (Chap. 12).

To create a muon neutrino beam, the decays of π^+, K^+ mesons are used. A high energy proton interacts with a target nucleon, producing many charged and neutral particles. A magnetic system (the *horn*) selects π^+, K^+ with a defined momentum (narrow-band neutrino beam) or with a large momentum range (broad-band neutrino beam).

In a narrow-band neutrino beam, π^+, K^+ are selected with momentum $p = 200$ GeV/c. The mesons are driven in a 1 km long vacuum tube, where they can decay. Determine:

(a) the π^+, K^+ mean free path;
(b) the fraction of π^+, K^+ decays at the end of the vacuum tube;
(c) the maximum energy in the laboratory system of the neutrinos produced in the π^+ and K^+ decays.
(d) Discuss the narrow-band neutrino beam flux shown in Fig. 8.10.

[See solutions]

10.10. **Neutrino beams-2. Contaminations**. Suppose that only positively charged particles are collected by the magnetic device (*horn*). The main source of neutrinos is the decay $\pi^+ \to \mu^+ \nu_\mu$. In this muon neutrino beam, there is an irreducible background of ν_e and $\overline{\nu}_\mu$.

(a) Write the processes which produce ν_e and $\overline{\nu}_\mu$ and give a simple estimate of their relative number with respect to the ν_μ as a function of the length L of the secondary particles decay tunnel.
(b) Evaluate the ν_e, $\overline{\nu}_\mu$ contamination for long baseline experiments (Chap. 12), assuming $L = 1$ km, $E_\mu = 10$ GeV.

[See solutions]

10.11. **Neutrino beams-3. Off-axis beams**. Assume that a neutrino beam is produced by the decay of high energy pions, with $E_\pi \gg m_\pi$, and that $E_\nu \gg m_\pi$.

(a) What is the characteristic angle θ_C of the decay neutrinos with respect to the direction of the pion in the lab. frame, when the neutrino is emitted at $\theta^* = 90°$ in the pion rest frame?
(b) What is the maximum angle $\theta_{max}(E_\nu)$ between the neutrino and the direction of its parent pion?
(c) What is the maximum energy $E_\nu(\theta)$ at which a neutrino can be produced in the decay of a pion if it appears at a given angle θ with respect to the pion direction?

Discuss the consequences of the existence of a maximum energy $E_\nu(\theta)$.

[See solutions]

Supplement 10.1: The Computing Effort at the LHC Collider

Data Reduction The Large Hadron Collider (LHC) experiments at CERN generate huge amounts of data, and special methods are therefore needed to analyze them efficiently. For a design luminosity of 10^{34} cm^{-2} s^{-1}, an average of 20 interactions per bunch crossing are expected every 25 ns, corresponding to an input rate of 40 MHz. The experiments, as for instance CMS, contain almost one hundred million detector channels. About 1 MB of zero-suppressed data would be required per event, resulting in 100 Terabyte of data per second. This is far too large to be handled and stored. Since most of the events are non interesting soft hadronic interactions, the trigger system performs the difficult task to preselect only interesting events. The challenge for the trigger system is therefore to reduce the input rate of 40 MHz by a factor $10^5 \div 10^6$ down to a rate of $\sim$100 Hz which is then written to a permanent storage. With an event size of about 1 MB, this corresponds to a data size rate of $\sim$100 MB/s, which is the maximum rate that can be sustained by the DAQ system to write and provide the data for the successive off-line reconstruction.

Due to the large complexity of LHC events the CMS experiment, for instance, has chosen to reduce this rate in two steps: a Level-1 Trigger and a High-Level Trigger.

The Level-1 trigger is purely hardware based. It is fast and performs the first rough estimates of relevant quantities. Custom synchronous processors which have access to a coarse granularity information from calorimeters and muon detectors are used to produce a list of roughly identified candidates for physics objects (such as jets o muons). It receives data at the full LHC bunch crossing rate of 40 MHz and takes the trigger decision for each bunch crossing within 3.2 μs. During this latency time, the full detector data are stored in front-end pipeline memories. The output rate is limited by the capabilities of the CMS data acquisition system to 100 kHz, corresponding to a first reduction factor of $\sim$400.

The High-Level Trigger is the second step of the trigger chain. It is designed to reduce the Level-1 output rate of $\sim$100 kHz to a final output rate of $\sim$100 Hz. The High-Level Trigger code runs on a farm of commercial processors and performs the reconstruction and selection of physics objects using the full event data with fine granularity and matching information from different sub-detectors.

The only way to reduce the overall analysis time to an acceptable level is to use multicore laptops, local clusters or global distributed clusters, i.e., the *Grid*. The challenge is to provide this *parallel computing* effort transparently so that users have not to worry about how to access all of these resources in parallel.

The Data GRID The Worldwide LHC Computing Grid [10w2] project is a global collaboration linking worldwide grid infrastructures and computer centers. It was launched in 2002 to provide global computing resources to store, distribute and analyze the 15 Petabytes (15 million Gigabytes) of data annually generated by LHC. The infrastructure, which serves a community of more than 8,000 physicists, is built by integrating thousands of computers and storage systems in hundreds of data centers worldwide; it enables a collaborative computing environment on a scale never seen before.

In 1999, when work on the design of a computing system for the LHC data analysis began, it rapidly became clear that the required computing power was far beyond the funding capacity available at CERN. Nevertheless, a solution was possible since most of the laboratories and universities across the world and that were collaborating to the LHC project, had access to national or regional computing facilities. The obvious question was: could these facilities be somehow integrated all together to provide a single LHC computing service?

During the development of the LHC Computing Grid, many additional benefits of a distributed "grid" system became evident:

- Multiple copies of data can be kept in different sites, ensuring access for all scientists involved, independently of their geographic location.
- Having computer centers in multiple time zones eases continuous monitoring and the availability of expert support.
- No single points of failure.
- Independently managed resources have encouraged novel approaches to computing and analysis.
- The system can be easily reconfigured to face new challenges, making it able to dynamically evolve throughout the LHC life span growing in capacity to meet the rising demands as more data is collected.
- The whole community can take advantage of new technologies that may appear and that offer improved usability, cost effectiveness or energy efficiency.
- Hardware is constantly being replaced and newer machines typically require recent operating systems and/or versions.

The LHC-grid project is made up of four layers, or "Tiers"; 0, 1, 2 and 3. Each tier provides a specific set of services.

Tier-0. It is the CERN Computer Center. All data from the LHC passes through this central hub, but it provides less than 20% of the total computing capacity. CERN is responsible for the safe-keeping of the raw data (first copy), first pass reconstruction, distribution of raw data and reconstruction output to the Tier-1's, and reprocessing of data during LHC down-times.

Tiers-1. These are large computer centers with sufficient storage capacity and with uninterrupted support for the Grid. They are responsible for the safe-keeping of raw and reconstructed data, large-scale reprocessing and safe-keeping of corresponding output, distribution of data to Tiers-2. The 11 Tiers-1 are located (managed) in Canada (TRIUMF), Germany (KIT), Spain (Port d'Informació Científica, PIC), France (IN2P3), Italy (INFN), Nordic countries (Nordic Datagrid Facility), Netherlands (NIKHEF/SARA), Taipei (ASGC), United Kingdom (GridPP) and USA (Fermilab-CMS; BNL-ATLAS).

Tiers-2. They are typically universities and other scientific institutes, which can store sufficient data and provide adequate computing power for specific analysis tasks. There are currently around 140 Tier-2 sites, covering most of the globe.

Tiers-3. Individual scientists can access these facilities through local computing resources, which can consist of local clusters in a University Department or even just an individual PC.

The ROOT Framework The LHC data are analyzed with object-oriented frameworks aimed at solving the data analysis challenges of high-energy physics. In recent years, the CERN standard program is ROOT [10w3]. ROOT offers an extensive and flexible set of tools for data storage, analysis and presentation. The data are defined as objects, and special storage methods are used to get direct access to the separate attributes of the selected objects, without having to touch the bulk of the data. The system was designed to query its databases in parallel on parallel-processing machines or on computer clusters. An example of result obtained with this framework is the full invariant mass spectrum of dimuons shown in Fig. 10.25.

Solutions

Problem 10.2 The energy-momentum four-vectors of the electron before and after the collision are respectively:

$$(E, \mathbf{p}); \qquad (E', \mathbf{p}')$$

The mass of the electron can be neglected compared to the total energy. Inserting the numerical values $E = 20$ GeV, $E' = 12$ GeV, $\theta = 5°$, the squared modulus of the transferred four-momentum $q = (E - E', \mathbf{p} - \mathbf{p}')$ is (see also Eq. (10.9)):

$$q^2 = -Q^2 = -4EE' \sin^2 \frac{\theta}{2} \longrightarrow Q^2 = 1.82\ \text{GeV}^2.$$

The relativistic invariant of the final state hadronic system W (see Fig. 10.5 and Eq. (10.29)) is:

$$W^2 = m_p^2 - Q^2 + 2m_p\nu = (0.938)^2 - (1.82)^2 + 15.0 = 14.1\ \text{GeV}^2$$

where $\nu = E - E' = 8$ GeV, $m_p = 0.938$ GeV, $2m_p\nu = 15.0$ GeV^2. It is clear that $W^2 > m_p^2$ and that the scattering is not elastic. The invariant mass of the formed hadronic system is:

$$W = \sqrt{14.1} = 3.75\ \text{GeV}$$

Problem 10.3 The integral of the momentum distribution of the u-type quark inside the proton over the variable x domain [0, 1] must be equal to 0.33:

$$0.33 = \int_0^1 a(1-x)^3 dx = \left[\frac{a}{4}(1-x)^4\right]_1^0 = \left[\frac{a}{4}\right]$$

from which $a = 1.32$ is obtained.

Problem 10.5 The approximate number of gluons in the interval $0.0001 < x < 0.001$ is given by (for $Q^2 = 10^4$ GeV^2):

$$N_g\left(10^{-4} - 10^{-3}\right) = \int_{0.0001}^{0.001} \frac{xg(x)}{x} dx = \int_{0.0001}^{0.001} 0.36\, x^{-1.5} dx \simeq 49.$$

Similarly $N_g(10^{-3} - 10^{-2}) \simeq 15$, $N_g(0.01 - 0.1) \simeq 5$, $N_g(0.1 - 1) \simeq 2$. A larger number of gluons is obtained for higher Q^2 and for smaller values of x.

Problem 10.6 In terms of quarks, the production is via the annihilations described in the reactions (8.64a), (8.64b) for the $W^{\pm}$ bosons and in the reaction (8.64c) for the Z^0 boson. The different quarks of the p carry a fraction of the momentum with respective distributions shown in Fig. 10.11. The antiquarks in the $\overline{p}$ have the same momentum distributions as that of the quarks in the p. In first approximation, the contribution of sea quarks can be neglected.

If E is the energy carried by the proton (antiproton), the energy carried by the x_i-type quarks is $x_i E$. As can be seen in Eq. (8.65), the W production cross-section has a maximum for a c.m. energy equal to the boson mass (~80 GeV). Thus, referring to Eq. (10.67), one has:

- 2 u-type (anti)quarks in p ($\overline{p}$) carrying on average 20% of the p ($\overline{p}$) momentum;
- 1 d-type (anti)quarks in p ($\overline{p}$) carrying on average 10% of the p ($\overline{p}$) momentum.

The c.m. energy of the system consisting of a quark in the proton and of an antiquark in the antiproton necessary to produce the reactions $u\overline{d} \to W^+$ Eq. (8.64a) and $d\overline{u} \to W^-$ Eq. (8.64b) is on average:

$$\sqrt{x_q x_{\overline{q}} E_p E_{\overline{p}}} = \sqrt{\frac{2(0.2) \times (0.1) + 1(0.2) \times (0.1)}{3}} \sqrt{s} = 0.14\sqrt{s}. \qquad (10.1)$$

At $\sqrt{s} = 630$ GeV, the relation (10.1) gives 88 GeV. Instead, at $\sqrt{s} = 540$ GeV, the available energy seems below threshold for the production of W vector bosons via quark-antiquark annihilations. Remember however that in Eq. (10.1), the average values were used: the quarks with energy in the tail of the distribution of Fig. 10.11 have enough energy to reach the production threshold.

A similar situation occurs for the boson Z^0, whose production cross-section is smaller.

It is easy to verify that the discovery would be far more difficult in a pp machine, where the vector boson production occurs through a valence quark and a sea antiquark. Note that the S$p\overline{p}$S collider was feasible thanks to the technological innovation (due to S. Van der Meer [10S85]) on the antiproton *stochastic cooling* allowing to obtain a high $\overline{p}$ luminosity.

Problem 10.7 The left-handed neutrinos and right-handed antineutrinos interact with left-handed quarks and right-handed antiquarks, as shown in Fig. 10.10. From this observation, the differential cross-sections for neutrinos and antineutrinos can be derived in terms of the variables x and y, see Eqs. (10.60) and (10.61):

$$\frac{d\sigma^{\nu}}{dxdy} = \sigma_0\big[xq(x) + x\overline{q}(x)(1-y)^2\big] \tag{10.2}$$

$$\frac{d\sigma^{\overline{\nu}}}{dxdy} = \sigma_0\big[xq(x)(1-y)^2 + x\overline{q}(x)\big] \tag{10.3}$$

Recalling that $\int_0^1 (1-y)^2 dy = 1/3$, the integral on the y variable of Eqs. (10.2) and (10.3) gives:

$$\frac{d\sigma^{\nu}}{dx} = \sigma_0\left[xq(x) + \frac{1}{3}x\overline{q}(x)\right] \tag{10.4}$$

$$\frac{d\sigma^{\overline{\nu}}}{dx} = \sigma_0\left[\frac{1}{3}xq(x) + x\overline{q}(x)\right] \tag{10.5}$$

By imposing that the ratio between Eq. (10.4) and Eq. (10.5) be equal to 0.5, one obtains:

$$\frac{d\sigma^{\overline{\nu}}/dx}{d\sigma^{\nu}/dx} = \frac{1/3xq(x) + x\overline{q}(x)}{xq(x) + 1/3x\overline{q}(x)} = 0.5 \longrightarrow xq(x) = 5x\overline{q}(x)$$

from which $q(x) = 5\overline{q}(x)$. The ratio between the momentum fraction carried by quarks with respect to that carried by antiquarks in isoscalar nuclei must be equal to 5 to explain the $\sigma^{\overline{\nu}}/\sigma^{\nu}$ ratio.

Problem 10.8 The value obtained in the previous problem is consistent with what can be obtained from the distributions of u, d quarks and $\overline{q}$ antiquark given in Eq. (10.67):

$$\int xu(x)dx = 0.2; \qquad \int xd(x)dx = 0.1; \qquad \int x\overline{q}(x)dx = 0.06$$

An isoscalar target with equal number of protons and neutrons have an equal number of u and d quarks. Therefore, one has:

$$\frac{\int x[u(x) + d(x)]dx}{\int x\overline{q}(x)dx} = \frac{0.2 + 0.1}{0.06} = 5$$

Problem 10.9

(a) The mean free path of a particle in the laboratory system is defined as $\lambda = \beta\gamma c\tau$, where τ is the particle lifetime in the rest frame. The π, K lifetimes are given in Table 7.3: $\tau_\pi = 2.6 \cdot 10^{-8}$ s, $\tau_K = 1.2 \cdot 10^{-8}$ s. Since $\beta\gamma = E/m$ and inserting numerical values ($m_\pi = 0.140$ GeV, $m_K = 0.494$ GeV), one finds:

$$\lambda_\pi = \frac{E}{m_\pi} c\tau_\pi = \frac{200}{0.14} \times 3 \cdot 10^{10} \times 2.6 \cdot 10^{-8} = 11.1 \text{ km}$$

$$\lambda_K = \frac{E}{m_K} c\tau_K = \frac{200}{0.494} \times 3 \cdot 10^{10} \times 1.2 \cdot 10^{-8} = 1.46 \text{ km}$$

(b) The fraction of decayed particles within the $L = 1$ km long vacuum tube is:

$$f_\pi = 1 - e^{-L/\lambda_\pi} = 8.6\%; \qquad f_K = 1 - e^{-L/\lambda_K} = 50.4\%$$

(c) To determine the energy of the neutrinos in the laboratory system, it is convenient to start from the c.m. system. If p_π, p_ν, p_μ are respectively the four-momenta of the initial state pion, and of the final state neutrino and muon, one has $p_\pi = p_\nu + p_\mu$. The corresponding relativistic invariant is:

$$(p_\pi - p_\nu)^2 = p_\mu^2 \longrightarrow m_\mu^2 = p_\pi^2 - 2p_\pi p_\nu \qquad (10.6)$$

Note that the last relation is obtained using the fact that the four-momentum squared is the particle rest mass: $p_\nu^2 = E_\nu^2 - |\mathbf{p}_\nu|^2 = m_\nu^2$. Therefore, for massless neutrinos, one has $p_\nu^2 = 0$.

Indicating with (*) the system in which the pion is at rest, one has $p_\pi^* = (m_\pi, 0)$ and from Eq. (10.6), one obtains:

$$m_\mu^2 = m_\pi^2 - 2m_\pi E_\nu^* \longrightarrow E_{\pi\to\nu}^* = \frac{m_\pi^2 - m_\mu^2}{2m_\pi} = 29.8 \text{ MeV} \qquad (10.7)$$

and similarly for the K decays:

$$E_{K\to\nu}^* = \frac{m_K^2 - m_\mu^2}{2m_K} = 236 \text{ MeV} \qquad (10.8)$$

The energy in the laboratory system is obtained by the Lorentz transformation:

$$E_\nu = \gamma\left(E_\nu^* + \beta p_\nu^* \cos\theta^*\right) \qquad (10.9)$$

where θ^* is the neutrino emission angle in the c.m. system with respect to the boost direction of the decaying particle. The maximum energy corresponds to the value $\theta^* = 0$. Recalling that $E_\nu = |\mathbf{p}_\nu|$ and that $\beta = 1$ (check with the values of γ obtained below), one obtains:

$$E_\nu = 2\gamma E_\nu^*$$

For 200 GeV/c pions, one has $\gamma = E_\pi / m_\pi = 200/0.140 = 1428$.
For 200 GeV/c kaons, one has $\gamma = E_K / m_K = 200/0.494 = 405$.
Therefore, the neutrino maximum energy, in the lab. system, is respectively for 200 GeV/c π and K decays:

$$E_{\pi\to\nu} = 2\gamma E^*_{\pi\to\nu} = 2\cdot 29.8\cdot 1428 = 85.1 \text{ GeV}$$
$$E_{K\to\nu} = 2\gamma E^*_{K\to\nu} = 2\cdot 236\cdot 404 = 191 \text{ GeV}$$

(c) The above result justifies the two-box structure visible in Fig. 8.10 for the narrow-band neutrino beam. Lower values of $E_{\pi\to\nu}$, $E_{K\to\nu}$ are obtained when the neutrino emission angle in the c.m. system is different from $\theta^* = 0$. As the pion/kaon spin is null, the θ^* angle distribution is isotropic and the corresponding distribution of the number of produced neutrinos as a function of the energy is flat up to the maximum value. The first step in Fig. 8.10 up to ~80 GeV is due to neutrinos from both $\pi\to\nu$ and $K\to\nu$ decays, while at higher energies only the K decays contribute.

Problem 10.10

(a) The ν_μ beam from the $\pi^+\to\mu^+\nu_\mu$ decay (which occurs with a probability $P(\pi\to\mu)$ in the decay tunnel) contains $\overline{\nu}_\mu$, ν_e from the subsequent decay $\mu^+\to e^+\nu_e\overline{\nu}_\mu$ (occurring with a probability $P(\mu\to e)$). The muon can decay only once produced, so the relative number of ν_e, $\overline{\nu}_\mu$ in a ν_μ beam is:

$$N \simeq \frac{P(\pi\to\mu)P(\mu\to e)}{P(\pi\to\mu)} = P(\mu\to e)$$

(we neglect the fact that the muon has an effective shorter $L - X$ decay length, with X the path travelled by the pion before decay). The decay probability for the muon is:

$$P(\mu\to e) = \left(1 - e^{-\frac{L}{\lambda_\mu}}\right) = \left(1 - e^{-\frac{Lm_\mu}{E_\mu c\tau_\mu}}\right) = \left(1 - e^{-0.16\frac{L\ [\text{km}]}{E_\mu\ [\text{GeV}]}}\right) \tag{10.10}$$

having used the fact that $\lambda_\mu = \frac{E_\mu}{m_\mu}c\tau_\mu$. In the last equality, the values of the constants (m_μ, c, τ_μ) were explicitly inserted.

Smaller contributions to the ν_e contamination arise from to the $\pi^+\to e^+\nu_e$ decay, with $BR \sim 10^{-4}$ and from the $K^+\to\pi^+\pi^0\nu_e$ decay, with $BR\sim 5\times 10^{-2}$. As the kaons produced by primary proton interactions are ~10% of the pions (see Table 10.2), the ν_e component from kaon decays is of the order of $\sim 0.1\times 0.05 = 5\times 10^{-3}$ that of the ν_μ.

(b) When the ratio $\frac{L\ [\text{km}]}{E_\mu\ [\text{GeV}]}$ in (10.10) is small, as in this case, the (10.10) can be approximate as:

$$P(\mu\to e) \simeq 0.16\frac{L\ [\text{km}]}{E_\mu\ [\text{GeV}]} = \frac{0.16\times 1}{10} = 1.6\%$$

having inserted the numerical values. The $\overline{\nu}_\mu$, ν_e contamination is of the order of the percent.

Problem 10.11

(a) The kinematics of the $\pi \to \mu\nu$ decay was derived in Problems 3.20 and 10.9. The pion has spin zero, so the neutrino emission angle θ^* follows an isotropic distribution in the pion rest frame. The relation for the angle θ in the lab. frame between the neutrino and its parent pion is the same as that obtained in Problem 3.20 for the muon (Eq. (3.33)):

$$\tan\theta = \frac{p_\perp}{p_\parallel} = \frac{p_\nu^* \sin\theta^*}{\gamma(\beta E_\nu^* + p_\nu^* \cos\theta^*)} = \frac{E_\nu^* \sin\theta^*}{\gamma E_\nu^*(\beta + \cos\theta^*)} \tag{10.11}$$

where $\gamma = 1/\sqrt{1-\beta^2} = E_\pi/m_\pi$ is the Lorentz boost. The characteristic angle of the decay in the lab. frame corresponds to $\theta^* = 90°$:

$$\tan\theta_C = \frac{1}{\gamma\beta} \longrightarrow \theta_C \simeq \frac{1}{\gamma} = \frac{m_\pi}{E_\pi} \ll 1 \tag{10.12}$$

when $E_\pi \gg m_\pi$.

(b) Let us consider the angle between the neutrino and its parent pion in the laboratory frame as a function on the neutrino energy rather than the pion energy. If $E_\pi > E_\nu \gg m_\pi$, then $\beta \sim 1$, $\gamma \gg 1$, Eq. (10.11) can be written as:

$$\tan\theta \simeq \frac{E_\nu^* \sin\theta^*}{\gamma E_\nu^*(\beta + \cos\theta^*)} = \frac{E_\nu^* \sin\theta^*}{E_\nu \cos\theta} \tag{10.13}$$

remembering the Lorentz transformation for the time-like component of the four-momentum, $p_\parallel = E_\nu \cos\theta = \gamma(\beta E_\nu^* + p_\nu^* \cos\theta^*)$. The maximum angle θ in the laboratory frame at which a neutrino of energy E_ν can appear is $\sin\theta = \frac{E_\nu^*}{E_\nu}\sin\theta^*$ and since $\sin\theta^* \leq 1$:

$$\theta_{max} \simeq \frac{E_\nu^*}{E_\nu} = \frac{29.8\text{ MeV}}{E_\nu} \tag{10.14}$$

which is small for most neutrinos. We used for E_ν^* the relation (10.7) found in Problem 10.9 for the pion decay.

(c) For a given angle θ, from Eq. (10.13), one obtains:

$$E_\nu(\theta) \simeq \frac{E_\nu^* \sin\theta^*}{\sin\theta} \leq \frac{29.8\text{ MeV}}{\sin\theta}\,. \tag{10.15}$$

For a given angle θ, there is a maximum possible neutrino energy $E_\nu(\theta)$. The existence of a maximum energy $E_\nu(\theta)$ for neutrinos that decay at a given angle with respect to their parent pions implies that many different pion energies contribute to this neutrino energy, which enhances the neutrino spectrum at this angle-energy combination.

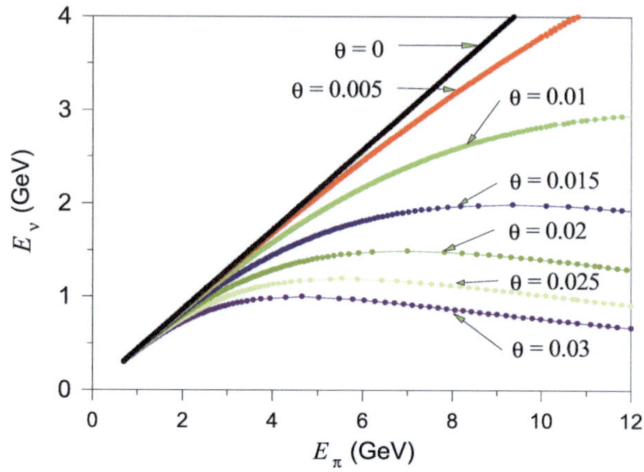

Fig. 10.1 Pion energy E_π needed to produce a neutrino of energy E_ν at various angles θ. For instance, the SuperKamiokande detector is $\theta \sim 2^\circ \sim 0.035$ off-axis the T2K beam. All neutrinos produced by the pion decay have energy below 1 GeV. At the distance of 295 km between Tokai (neutrino production) and Kamioka (neutrino detection), a maximal neutrino oscillation is expected to occur at energies lower than 1 GeV

If, for particular reasons, one experimental apparatus needs an enhancement at a particular neutrino energy,[1] the experiment must be displaced by the angle θ_{max}, Eq. (10.14), with respect to the direction of the neutrino beam. The value of θ_{max} is independent of the proton/pion energy. In this case, the neutrino beam is called *off-axis*. The neutrino energy spectrum produced by a typical pion spectrum can be derived using Monte Carlo simulations. An example is shown in Fig. 10.1 [10M01].

References

[10S85] van der Meer, S.: Stochastic cooling and the accumulation of antiprotons. Rev. Mod. Phys. **57**, 689–697 (1985)

[10M01] McDonald, K.T.: An off-axis neutrino beam. arXiv:hep-ex/0111033

[10T11] Abe, K., et al.: The T2K Experiment. Nucl. Instrum. Methods Phys. Res., Sect. A, Accel. Spectrom. Detect. Assoc. Equip. **659**, 106–135 (2011). Also on arXiv:1106.1238

[10w2] http://lcg.web.cern.ch/lcg/public/default.htm

[10w3] ROOT: a data analysis framework. http://root.cern.ch

[10w4] http://www-off-axis.fnal.gov/

[1] This is the case of proposed NOνA experiment at the NuMI neutrino beam [10w4] in US and the running T2K experiment in Japan [10T11]. These experiments are devoted to the study of $\nu_\mu \to \nu_e$ oscillations and to the measurement of the neutrino mixing angle θ_{13}, see Chap. 12.

Chapter 11
The Standard Model of the Microcosm

Problems

11.1. **Lagrangian density for the Klein-Gordon equation**. In a quantum field theory invariant under Lorentz transformations, the fields ϕ_i are functions of the space-time coordinates x_μ and the Lagrange equation (6.1) is written in terms of the Lagrangian density $\mathcal{L}(\phi_i, \partial_\mu \phi_i)$:

$$\frac{\partial \mathcal{L}}{\partial \phi_i} - \frac{\partial}{\partial x_\mu} \frac{\partial \mathcal{L}}{\partial(\partial_\mu \phi_i)} = 0 \tag{11.1}$$

such that $L = \int \mathcal{L} d^3x$. Show that, for a scalar field ϕ describing spinless particles of mass m and Lagrangian density:

$$\mathcal{L}_\phi = \frac{1}{2}\partial^\mu \phi \partial_\mu \phi - \frac{1}{2} m^2 \phi^2, \tag{11.2}$$

the Klein-Gordon equation (4.13) is obtained.
[See solutions]

11.2. **Lagrangian density for the Dirac equation**. Similarly to the previous problem, show that for a field ψ describing spin 1/2 fermions with a Lagrangian density:

$$\mathcal{L}_\psi = i\overline{\psi}\gamma^\mu \partial_\mu \psi - m\overline{\psi}\psi, \tag{11.3}$$

the Dirac equation (4.16) is obtained.
[See solutions]

11.3. **Z^0 partial width**. Calculate the value predicted by the electroweak model for the partial width Γ_{ν_e} for the decay of a Z^0 boson in a $\nu_e, \overline{\nu}_e$ pair at the energy corresponding to the peak of the Z^0 resonance.
[See solutions]

S. Braibant et al., *Particles and Fundamental Interactions: Supplements, Problems and Solutions*, Undergraduate Lecture Notes in Physics,
DOI 10.1007/978-94-007-4135-5_11,

11.4. **The running α_S**. Using Eq. (11.90):

$$\alpha_S(Q^2) = \frac{12\pi}{(33-2N_f)\ln(\frac{Q^2}{\Lambda^2_{QCD}})} \tag{11.4}$$

compute $\alpha_S(Q^2)$ for (a) $Q = 8$ GeV, (b) $Q = 90$ GeV and (c) $Q = 500$ GeV. For each energy, the number N_f of active flavors must be considered. Assume $\Lambda_{QCD} = 200$ MeV (actually Λ_{QCD} varies discontinuously when it exceeds the threshold corresponding to the mass of a new quark).
[See solutions]

11.5. **α_s at the charm threshold**. Compare the result of the calculation of the α_S constant obtained using Eq. (11.4) at energies corresponding to the *charmonium* production, with the result obtained in Problem 9.1.
[See solutions]

11.6. **Non-abelian theory**. Explain the meaning of a non-abelian theory from a physical point of view.
[See solutions]

11.7. **SU(2) and U(1) symmetry groups**. Explain the meaning of the SU(2) and U(1) groups. Say whether they are or not abelian groups (for the definition of abelian group, see the solution of the previous problem).
[See solutions]

Solutions

Problem 11.1 The derivatives of the Lagrangian density $\mathcal{L}_\phi$ with respect to ϕ and $\partial_\mu\phi$ are:

$$\frac{\partial\mathcal{L}_\phi}{\partial\phi} = -m^2\phi; \qquad \frac{\partial\mathcal{L}_\phi}{\partial(\partial_\mu\phi)} = \partial^\mu\phi.$$

From these above derivatives and using Eq. (11.1), one obtains:

$$\partial^\mu\partial_\mu\phi + m^2\phi = 0$$

which is the Klein-Gordon equation in natural units ($\hbar = c = 1$).

Problem 11.2 The derivative of the Lagrangian density $\mathcal{L}_\psi$ with respect to ψ and $\partial_\mu\psi$ are:

$$\frac{\partial\mathcal{L}_\psi}{\partial\psi} = -m\overline{\psi}; \qquad \frac{\partial\mathcal{L}_\psi}{\partial(\partial_\mu\psi)} = i\overline{\psi}\gamma^\mu$$

from which, one obtains:

$$-i\partial_\mu \overline{\psi}\gamma^\mu - m\overline{\psi} = 0$$

and its conjugate equation:

$$i\gamma^\mu \partial_\mu \psi - m\psi = 0$$

which corresponds to the Dirac equation (4.16) in natural units.

Problem 11.3 The Z^0 width is given in Eq. (9.19):

$$\Gamma_\nu = N_c^f \frac{G_F m_Z^3}{6\sqrt{2}}(v_f^2 + a_f^2)$$

where the color factor $N_c^f = 1$ for leptons. The couplings v_f, a_f for different fermions are calculated in the framework of the Standard Model and are reported in Table 11.3. For neutrinos: $v_f = a_f = 0.5$. Inserting the numerical values, one obtains:

$$\Gamma_\nu = \frac{1.17\ 10^{-5}\ [\text{GeV}]^{-2} \times (91.2\ [\text{GeV}])^3 \times (0.5^2 + 0.5^2)}{6\sqrt{2}\pi} = 0.166\ \text{GeV}$$

to be compared with the measured value reported in Table 9.4.

Problem 11.4 The value $Q = 8$ GeV is below the threshold for the production of the t quark, then the number of active flavors is $N_f = 5$. In this case $\alpha_S(Q = 8\ \text{GeV}) = \frac{37.7}{(33-10)\cdot 7.4} = 0.22$.
The value $Q = 90$ GeV is below the threshold for the production of the t quark, and $N_f = 5$. Then $\alpha_S(Q = 90\ \text{GeV}) = 0.13$.
Finally, at $Q = 500$ GeV, $N_f = 6$. Then $\alpha_S(Q = 500\ \text{GeV}) = 0.11$.

Problem 11.5 In this case, $Q \sim 3$ GeV and N_f=4 and $\alpha_S(Q = 1.5\ \text{GeV}) = 0.28$, to be compared with the value $\alpha_S \sim 1$ obtained in Problem 9.1.

Problem 11.6 The term *non-abelian* comes from the group theory: "non-abelian" means that $A \times B \neq B \times A$ and involves the matrices algebra [A89]. The physical meaning can be better understood in terms of Feynman diagrams.

QED is an abelian theory. In terms of Feynman diagrams, this means that a diagram with 3 or more photons meeting in a point does not exist: photons do not interact directly with each other. The interaction between photons can only take place through diagrams of the type shown in Fig. 11.13.

QCD is a non-abelian theory. The QCD gluons interact with each other and give rise to diagrams where 3 or 4 gluons meet in one point, see Fig. 11.12.

The electroweak theory is non-abelian. A photon can interact directly with a W^+W^- pair and similarly a Z^0 interacts directly with a W^+W^- pair.

Problem 11.7 SU(2) is represented by 2×2 matrices of the type

$$\begin{pmatrix} a & b \\ c & d \end{pmatrix}$$

where a, b, c and d can be complex numbers such that:

$$a^* = -d; \qquad b^* = -c; \qquad c^* = -b$$
$$ad - bc = 1$$

For example, the group of rotations in the three-dimensional space of spin $1/2$ particles is included in this representation. Such a group can be generated by the Pauli matrices. U(1) corresponds to e^{ix} where x is a real number. U(1) is an abelian group, while SU(2) is not.

Chapter 12
CP-Violation and Particle Oscillations

Problems

12.1. **Neutron-antineutron mixing**. Neutron and antineutron are each the antiparticle of the other; they are neutral, just as the K^0 and $\overline{K^0}$. Explain why a mixing between K^0 and $\overline{K^0}$ can occur, while the mixing between neutron and antineutron is forbidden.
[See solutions]

12.2. $\boldsymbol{K^0, \overline{K}^0}$ **mixing-1**. Describe how to produce a pure K^0 meson beam. The initial K^0 beam evolves during propagation in a mixed K^0 and $\overline{K^0}$ state (see Sect. 12.2). The mass difference between the two mass eigenstates K_1^0 and K_2^0 is $\Delta m = m_2 - m_1 \simeq 1/\tau_1$, with $\tau_1 = 90 \times 10^{-12}$ s.
(a) Evaluate Δm;
(b) Write the K^0 and $\overline{K^0}$ intensities as a function of proper time;
(c) Draw a graph of the intensity of the K^0 and $\overline{K^0}$ beams as a function of the proper time τ_1.
[A: See Sect. 12.2.1]

12.3. $\boldsymbol{K^0, \overline{K}^0}$ **mixing-2**. A K^0 beam propagating in vacuum can decay. At a distance d corresponding to 20 times the K_1 lifetime ($d = 20c\tau_{K_1}$) there is a target that absorbs 10% of the incoming K^0 beam. If the interaction cross-section for $\overline{K^0}$ is three times larger than that of the K^0, calculate the relative amplitudes of K_1 and K_2 in the beam:
(a) At $t = 0$;
(b) Immediately before the target;
(c) Immediately after the target.
Assume low-energy kaons, and neglect relativistic effects.
[See solutions]

S. Braibant et al., *Particles and Fundamental Interactions: Supplements, Problems and Solutions*, Undergraduate Lecture Notes in Physics,
DOI 10.1007/978-94-007-4135-5_12,

12.4. **$\varepsilon, \varepsilon'$ CP-violation parameters**. The quantities η_{+-}, η_{00}, defined in Eqs. (12.16) and (12.17), are related to the CP-violation parameters ε and ε' through the relations:

$$\eta_{+-} = |\eta_{+-}|e^{i\varphi_{+-}} = \varepsilon + \varepsilon' \tag{12.1}$$

$$\eta_{00} = |\eta_{00}|e^{i\varphi_{00}} = \varepsilon - 2\varepsilon' \tag{12.2}$$

Demonstrate that the ratio ε'/ε can be determined by measuring the double ratio R:

$$R = \frac{|\eta_{00}|^2}{|\eta_{+-}|^2} = \frac{\Gamma(K_L \to \pi^0\pi^0)}{\Gamma(K_S \to \pi^0\pi^0)} \bigg/ \frac{\Gamma(K_L \to \pi^+\pi^-)}{\Gamma(K_S \to \pi^+\pi^-)} \simeq 1 - 6\frac{\varepsilon'}{\varepsilon}. \tag{12.3}$$

[See solutions]

12.5. **K_L semi-leptonic decay.** Demonstrate that in the K_L semi-leptonic decays, the asymmetry:

$$A_L = \frac{\Gamma(K_L \to \pi^-\ell^+\nu_l) - \Gamma(K_L \to \pi^+\ell^-\overline{\nu}_l)}{\Gamma(K_L \to \pi^-\ell^+\nu_\ell) + \Gamma(K_L \to \pi^+\ell^-\overline{\nu}_\ell)}. \tag{12.4}$$

(ℓ indicates the muon or the electron) is related to the CP-violation parameter ε through the relation:

$$A_L \simeq 2\,\mathrm{Re}(\varepsilon) = (3.32 \pm 0.06) \cdot 10^{-3} \tag{12.5}$$

Show that the experimental result of (12.5) is consistent with that obtained in Eqs. (12.23–12.24) for non-leptonic decays.
[See solutions]

12.6. **B^0 meson tagging.** The B^0 meson *tagging*, as shown in Fig. 12.8, relies on the presence of a μ^+ amongst the decay products. The branching ratio for the decay into a positive charged lepton is $BR(B^0 \to \mu^+\nu_\mu + anything) =$ 10.3%. The decay into $e^+\nu_e$ has the same BR.

Explain the reason why a positive charged lepton is expected from the B^0 decay with the quoted BR.
[See solutions]

12.7. **Solar neutrino detection.** An experiment devoted to detect neutrinos produced by the Sun is carried out in a mine using the reaction

$$\nu_e + {}^{37}\mathrm{Cl} \to {}^{37}\mathrm{Ar} + e^- \tag{12.6}$$

The detector contains v $\simeq 4 \cdot 10^5$ liters of tetra-chloroethylene (C_2Cl_4). Estimate the number of ^{37}Ar atoms that would be produced per day using the following assumptions:

1. The solar luminosity is (see Appendix A5) $L_\odot = 3.84 \cdot 10^{26}$ W;
2. 7% of the Sun thermonuclear energy is emitted as neutrinos of average energy $\langle E \rangle \simeq 1$ MeV;
3. Only 0.1% of these neutrinos have enough energy to trigger the $(\nu_e + \text{Cl})$ reaction;
4. The interaction cross-section, for neutrinos able to induce the reaction on the ^{37}Cl nucleus, is $\sigma = 10^{-43}$ cm^2;
5. The isotopic abundance of ^{37}Cl is 25%;
6. The density of C_2Cl_4 is $\rho = 1.5$ g cm^{-3} and its molecular weight is $P = 164$ g mole^{-1}.

[See solutions]

12.8. **Atmospheric neutrino oscillation-1**. Using the two flavor oscillation formula (12.49), determine the value of the muon neutrino energy E_ν which gives a 100% disappearance probability for:
(a) neutrinos crossing the Earth atmosphere ($L \sim 20$ km);
(b) neutrinos crossing the Earth diameter ($L \sim 13000$ km).
Comment the result. Use the best fit value of Δm^2 and $\sin^2 2\theta$.
[A: (a) 40 MeV, below the muon production threshold; (b) 23 GeV]

12.9. **Atmospheric neutrino oscillation-2**. The MACRO [12A98] experiment observed atmospheric muon neutrino from below. The neutrinos, interacting below the detector, produce upgoing muons whose zenith direction θ was measured. The zenith angle is correlated with the neutrino path length L through the relation:

$$L = R_T \cos\theta + \sqrt{(R_T \cos\theta)^2 + 2R_T h + h^2} \tag{12.7}$$

where $R_T = 6370$ km is the Earth radius and $h \simeq 20$ km the atmospheric depth. A vertical upgoing muon has $\theta = 0$, while a downward going has $\theta = \pi$.
(a) Draw the oscillation probability for $\nu_\mu \to \nu_\tau$ as a function of $\cos\theta$ for $E_\nu = 20, 50, 100, 200$ GeV;
(b) Assuming that the MACRO signal from the vertical direction is induced by ν_μ with energy between 5 and 100 GeV, evaluate the survival probability $P(\nu_\mu \to \nu_\mu)$
[See solutions]

12.10. **Atmospheric neutrino oscillation-3**. Atmospheric neutrinos contain both $\nu_\mu, \overline{\nu}_\mu$ from $\pi^\pm$, $K^\pm$ decays. Evaluate the ratio $R = (\nu_e + \overline{\nu}_e)/(\nu_\mu + \nu_\mu)$ for atmospheric neutrinos assuming $L = 20$ km and for $E_\mu = 0.2, 20, 200$ GeV.

The measurement of an anomaly in the ratio R at low energy was, at the end of the 1980s, the first indication of an anomaly in the atmospheric neutrinos, which led to the discovery of the neutrino oscillations [12K04].
[See solutions]

12.11. **Neutrino oscillation matrix**. Show that the 3×3 unitary matrix (12.51) representing the mixing between flavor and mass eigenstates can be parameterized as:

$$U_{fj} = \begin{pmatrix} 1 & 0 & 0 \\ 0 & c_{23} & s_{23} \\ 0 & -s_{23} & c_{23} \end{pmatrix} \begin{pmatrix} c_{13} & 0 & s_{13}e^{-i\delta} \\ 0 & 1 & 0 \\ -s_{13}e^{i\delta} & 0 & c_{13} \end{pmatrix} \begin{pmatrix} c_{12} & s_{12} & 0 \\ -s_{12} & c_{12} & 0 \\ 0 & 0 & 1 \end{pmatrix} \tag{12.8}$$

Remember that the symbol f represents one of the *flavor eigenstate* ($f = e, \mu, \tau$). The symbol j stands for one of the *mass eigenstate* ($j = 1, 2, 3$). Thus, e.g., the matrix element $U_{\mu 3}$ in (12.51) is:

$$U_{\mu 3} = s_{23}c_{13} = \sin\theta_{23}\cos\theta_{23}$$

12.12. **Measurement of the small θ_{13} mixing angle**.
(a) Using the neutrino mixing matrix given in (12.51), show that for long baseline experiments (as well as for the atmospheric neutrinos), the contribution from Δm^2_{12} terms can be neglected and that the oscillation probabilities can be expressed in a simplified manner as:

$$P(\nu_\mu \to \nu_e) = \sin^2 2\theta_{13} \sin^2\theta_{23} \sin^2 \Phi_{23} \tag{12.9a}$$

$$P(\nu_\mu \to \nu_\tau) = \sin^2 2\theta_{23} \cos^4\theta_{13} \sin^2 \Phi_{23} \tag{12.9b}$$

$$\begin{aligned} P(\nu_\mu \to \nu_\mu) &= 1 - P(\nu_\mu \to \nu_e) - P(\nu_\mu \to \nu_\tau) \\ &= 1 - \sin^2 2\theta_{13} \sin^2\theta_{23} \sin^2 \Phi_{23} \\ &\quad - \sin^2 2\theta_{23} \cos^4\theta_{13} \sin^2 \Phi_{23} \end{aligned} \tag{12.9c}$$

$$P(\nu_e \to \nu_\tau) = \sin^2 2\theta_{13} \cos^2\theta_{23} \sin^2 \Phi_{23} \tag{12.9d}$$

$$P(\nu_e \to \nu_e) = 1 - \sin^2 2\theta_{13} \sin^2 \Phi_{23} \tag{12.9e}$$

where $\Phi_{23} = \Delta m^2_{23} L/4E_\nu \simeq \Delta m^2_{13} L/4E_\nu$.
(b) Describe the best way to experimentally measure θ_{13} with present technology.
[See solutions]

Supplement 12.1: Analogy for the Neutrino Mixing

To clarify the concept of neutrino mixing, we consider an orthogonal Cartesian coordinate system whose x and y axes correspond to the flavor eigenstates ν_μ and ν_τ. A second orthogonal Cartesian system, with coordinates x', y' corresponding to the mass eigenstates ν_2 and ν_3, is rotated by an angle θ with respect to the x, y system. A point $P(\nu_\mu, 0)$ on the x axis is a *pure flavor* eigenstate ν_μ; the corresponding point $P(x'_0, y'_0)$ in the second reference system has a component along x' and a component along y' and is a mixture of two mass eigenstates. The amount of mixing is determined by the rotation angle θ of the second system with respect to the first, the so-called *mixing angle*. If the mixing angle is small, the mass eigenstates are almost pure flavor eigenstates and vice versa. The energy associated to the mass eigenstates increases as their rest mass increases (for the same momentum p). According to the mass-energy equivalence, the rest mass of a particle contributes to its total energy. Like all particles during propagation, the neutrino mass eigenstates are represented by waves whose frequency increases with energy. Following the atmospheric ν_μ path (or that of the solar ν_e to the Earth), the neutrino mass eigenstates are waves propagating with different frequencies. If the mass eigenstates have the same eigenvalue (degenerate states), the waves will reach the detector with the same phases. These waves *recombine* to give back exactly the expected ν_μ or ν_e eigenstate detected via a weak interaction in an experimental apparatus. If the mass eigenstates have different eigenvalues, the waves propagate with different frequencies; they reach the detector with different phases. The waves, as they recombine, do not always reproduce the initial flavor eigenstate (for instance, ν_e for solar neutrinos).

Supplement 12.2: Dirac or Majorana Neutrinos: the Double β Decay

In its original formulation, the Standard Model assumes that neutral elementary fermions (the neutrinos) are massless. This allows the flavor lepton conservation without introducing additional parameters in the model [12A08].

In the most conventional scheme, neutrinos are *Dirac particles*: neutrinos can be distinguished from antineutrinos through their interactions. But parity violation tells us that neutrinos are *left-handed* and antineutrinos *right-handed*. In the four-component representation of Dirac fermions of Appendix 4, only two states are non vanishing. Therefore, there is no reason to exclude the possibility that massless neutral fermions be the components of a doublet (*Majorana particles*) in which "neutrinos" coincide with their own antiparticles: the *neutrino* is the left-handed component of the doublet while the *antineutrino* is the right-handed one.

Weak interactions favor the coupling of left-handed "neutrinos" with negative leptons and that of right-handed "neutrinos" with positive leptons. The sequence of reactions:

$$n \to p + e^- + \overline{\nu}_e \quad \text{is observed} \tag{12.10a}$$

$$\overline{\nu}_e + p \to n + e^+ \quad \text{is not observed} \tag{12.10b}$$

The process (12.10b) is not observed even if the reaction $\overline{\nu}_e + p \to n + e^+$ exists in nature. This does not imply that ν_e and $\overline{\nu}_e$ are different objects, because the weak interactions have a $V - A$ structure, see Sect. 8.16. The particles which we call

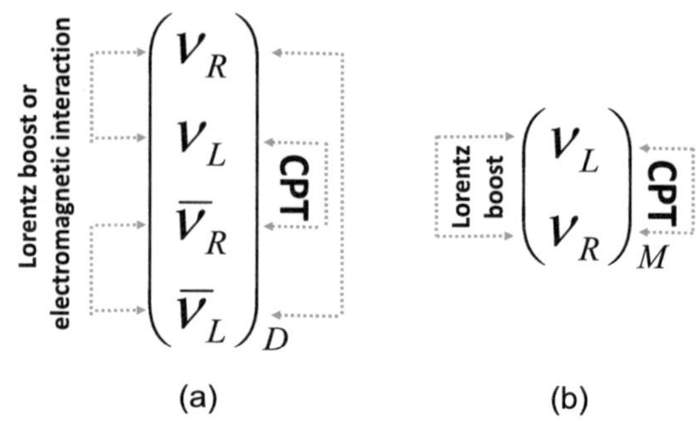

Fig. 12.1 Transformations connecting massive Dirac (**a**) and Majorana (**b**) neutrinos

$\overline{\nu}_e$ are right-handed and therefore the non observation of the process (12.10b) could simply be due to a strong dynamical suppression because of the wrong helicity state of the incoming "neutrinos". This helicity suppression is the same $P_{LH} = (1 - v/c)$ factor as that obtained in Sect. 8.10 for the pion decay. For the neutrino, since $v/c = p_\nu/E_\nu$, one can write:

$$\left(1 - \frac{v}{c}\right) = \left(1 - \frac{p_\nu}{E_\nu}\right) = 1 - \frac{\sqrt{E_\nu^2 - m_\nu^2}}{E_\nu} \simeq 1 - \frac{E_\nu\left(1 - \frac{m_\nu^2}{2E_\nu^2}\right)}{E_\nu} = \left(\frac{m_\nu}{2E_\nu}\right)^2 \quad (12.11)$$

The sequence of reactions described in (12.10a) and (12.10b) is therefore possible if two conditions are met:

- ν_e and $\overline{\nu}_e$ are identical particles (that is the "neutrino" is a Majorana particle)
- The neutrino mass m_ν is non vanishing, so that the helicity suppression is large but not complete.

Following the reference [12K89], it is possible to test the nature of the neutrino (Dirac or Majorana particle) only if the neutrino is not massless (as found after the discovery of neutrino oscillations). To understand the difference between Dirac and Majorana particles, let us consider the existence of a massive left-handed neutrino ν_L and the transformations shown in Fig. 12.1. Assuming the CPT invariance to be valid, the existence of ν_L implies the existence of the CPT mirror image: a right-handed antineutrino $\overline{\nu}_R$. In addition, as a massive ν_L travels slower than light, a reference frame moving faster than ν_L can be found: in this reference frame, the neutrino travels in the opposite direction but its spin is unchanged. Therefore, the Lorentz transformation to this faster moving reference frame transforms ν_L into a right-handed ν_R. This ν_R may be or may not be the same particle as the CPT mirror image $\overline{\nu}_R$ of ν_L. If it is not the same, ν_R has its own CPT mirror image $\overline{\nu}_L$. Altogether, there are four states with the same mass (see Fig. 12.1a). This quadruplet of state is a Dirac neutrino denoted as ν_D. It has distinct particle and antiparticle states. Dirac neutrinos may also have a magnetic dipole moment and be subjected to the electromagnetic interaction. In summary, in presence of a mass, Dirac neutrinos can transform one into the other through CPT, Lorentz boost and electromagnetic interactions. On the other hand, if the right-handed particle obtained by the

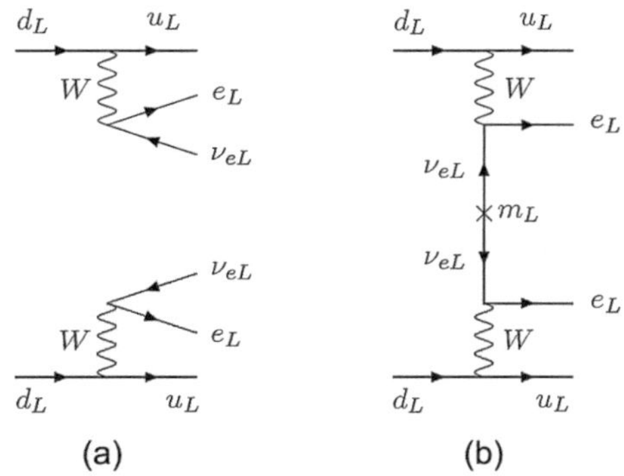

Fig. 12.2 Double β decay diagrams. Diagram (**a**) shows the standard $2\nu\beta^-\beta^-$ process in which two neutrinos are emitted; diagram (**b**) shows the neutrinoless decay. This process is possible only if the neutrino is a massive Majorana particle; it is searched for in different experiments but has not yet been observed

Lorentz transformation to the faster moving reference frame, is the same particle as the CPT image of the original ν_L, there are only two states with a common mass (see Fig. 12.1b) differing only through their helicity. This state doublet represents a Majorana neutrino denoted as ν_M. There is no need to define a *neutrino* and a *antineutrino* state. If the "neutrino" is a Majorana particle, the **lepton number conservation** is violated. Note that Majorana neutrinos can transform one into the other only by Lorentz boost and CPT. Indeed, under CPT, the fields **E** and **B** remain unchanged while the spin of a Majorana neutrino reverses its direction. Therefore, if the CPT invariance holds, the dipole moments must vanish and an electromagnetic interaction is not possible.

The most promising way to distinguish between Dirac and Majorana neutrinos is the neutrinoless double β decay [12K00]. The double β decay processes are described by the following reactions:

$$(Z, A) \to (Z+2, A) + 2e^- + 2\overline{\nu}_e \quad \left(2\overline{\nu}\beta^-\beta^-\right) \tag{12.12a}$$

$$(Z, A) \to (Z-2, A) + 2e^+ + 2\nu_e \quad \left(2\nu\beta^+\beta^+\right) \tag{12.12b}$$

that can occur when a single β decay is kinematically forbidden. For example (see Fig. 14.11), the nucleus ^{106}Cd ($Z = 48$) cannot have a β decay into the $Z = 47$ state ^{106}Ag that has a mass 0.2 MeV larger, but can have a double β decay into the $Z = 46$ state ^{106}Pd that is 2.8 MeV lighter. The process (12.12a) at the fundamental (quark) level, shown in Fig. 12.2a, is the transition of two left-handed d quarks into two left-handed u quarks with the emission of two electrons and two $\overline{\nu}_e$. The process is of second order in the weak coupling and therefore, the corresponding decay rates are very low and with lifetimes of the order of $t_{1/2} > 10^{19}$–10^{21} years.

On the contrary, the *neutrinoless double β decay* is the exotic process:

$$(Z, A) \to (Z+2, A) + 2e^- \quad (0\nu\beta\beta) \tag{12.13}$$

without neutrino emission. The leading order diagram of this process is shown in Fig. 12.2b and can be pictured as a β decay followed by the absorption of the emitted antineutrino by a different neutron in the nucleus. The process has a very clear

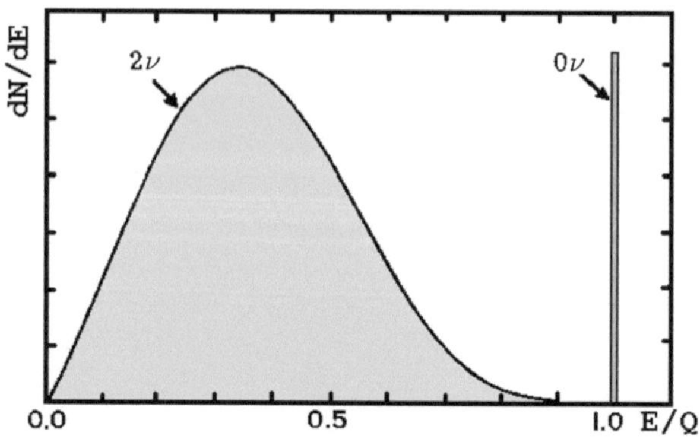

Fig. 12.3 Energy spectra of the ($2\nu\beta\beta$) and ($0\nu\beta\beta$) decays. The energy scale is normalized to the maximum available kinetic energy Q

experimental signature; indeed, in the standard decay, the sum of the energy of the two final state electrons has a broad distribution, while in the neutrinoless case, the sum of the energies of the two emitted electrons is equal to the Q value of the reaction, where Q is the difference between the masses of the initial and final nuclei (see Fig. 12.3).

In order to have a neutrinoless double β decay, the neutrino must be its own antiparticle, that is it must be a Majorana particle, and the neutrino mass must be non zero. The first condition can be easily understood observing that the virtual neutrino in the diagram of Fig. 12.2b is emitted as $\overline{\nu}_e$ while it is absorbed as a ν_e. Reaction (12.13) clearly violates the lepton number by two units. The second condition results from the fact that the antineutrino emitted in the "first" decay has the wrong helicity for being absorbed. The absorption is therefore possible without violating angular momentum conservation only if $m_\nu > 0$. It can be shown that the amplitude for neutrinoless double β decay is proportional to the neutrino mass through the relation:

$$M(0\nu\beta\beta) \propto \sum_{i=1,3} U_{ei}^2 m_i = \langle m_{\nu_e} \rangle_{eff} \tag{12.14}$$

where U is the neutrino mixing matrix. The effective mass $\langle m_{\nu_e} \rangle_{eff}$ can be smaller than the smallest neutrino mass m_i. The quantity $\langle m_{\nu_e} \rangle_{eff}$ constitutes a lower limit on the mass of the heaviest neutrino.

More than 60 naturally occurring isotopes fulfill condition for $2\nu\beta\beta$. Only eleven isotopes have been experimentally observed undergoing two neutrino double β decay: ^{48}Ca, ^{76}Ge, ^{82}Se, ^{96}Zr, ^{100}Mo, ^{116}Cd, ^{128}Te, ^{130}Te, ^{130}Ba, ^{150}Nd, and ^{238}U. Other isotopes can be subjected to both double β decay and other decays. In most cases, the double β decay lifetime is so long that it is nearly impossible to disentangle a signal against the background from other radiation.

Double β decays are searched for with different experimental techniques. Direct detection in a sample material occurs if the decay is recorded in real time. Since decays can be detected event by event, energy and (in principle) angular distribution of the final state e^- or e^+ can be reconstructed if tracks of the charged particles can be reconstructed (*tracking experiments*). The charge sign can be determined if a magnetic field is applied along the path of the particles. More frequently, the total energy spectrum of the decay electrons (or positrons) is determined in a calorimetric approach (*calorimetric experiments*). The occurrence of a two neutrino or a neutrinoless double β decay is identified according to the classification shown in Fig. 12.3.

The current best limits on neutrinoless double β decay come from calorimetric experiments located in the Gran Sasso underground laboratory using ^{76}Ge and ^{130}Te [12C11]. All experiments require a very low background (of the order of 0.2 counts/year/kg/keV) at the expected position of the $0\nu\beta\beta$ signal. The half-live limit for the neutrinoless mode is of the order of $t_{1/2} \geq 10^{25}$ years. This limit can be used to extract a limit on the effective mass (see Eq. (12.14)) at 90% C.L.

$$\langle m_{\nu_e} \rangle_{eff} \leq 0.35 \text{ eV} \tag{12.15}$$

the effective mass parameter from the measured lifetime limit is derived after a difficult calculation of the matrix element between the initial and final nuclei. This introduces some uncertainty in the limit.

Solutions

Problem 12.1 The oscillations of the $K^0, \overline{K}^0$ system is due to a second order weak interaction transition between two quarks, $d \to s$ and between two antiquarks $\overline{s} \to \overline{d}$ (see Fig. 12.3). In this case, a quark transforms into a different flavor quark and an antiquark into another antiquark.

The transition neutron $\leftrightarrow$ antineutron requires the transformation of three quarks into three antiquarks. But any quark-antiquark transformation is forbidden.

Problem 12.3

(a) The production of K^0 occurs through the reaction (12.1). Referring to Fig. 12.1, the initial K^0 beam is composed of 50% of K_1 and of 50% of K_2, Eq. (12.6b).

(b) The target is located at a distance $d = 20c\tau_{K_1} = 54$ cm. Following the radioactive decay law given in Eq. (4.42), the K_1 component in the beam is attenuated by a factor $\frac{N(t)}{N_0} = e^{-t/\tau} = e^{-20\tau/\tau} = e^{-20} \simeq 2 \cdot 10^{-9}$. At this distance, all K_1 have disappeared. Since the K_2 component lifetime is about 600 times longer than that of the K_1 (see Sect. 12.2.1, $\tau_{K_2} \simeq 600\tau_{K_1}$), its attenuation is negligible. Therefore, before the target, the beam contains almost exclusively the K_2 component.

(c) The K_2 beam is in turn composed of 50% of K^0 and of 50% of $\overline{K}^0$. The eigenstates interacting via strong interaction when the beam passes through the material (assumed of negligible thickness compared to the K_1 lifetime) are the K^0

and $\overline{K}^0$. These have different interaction probabilities: 10% of the K^0 beam is attenuated (e.g., from 50 initial K^0, 45 survive). Since the interaction cross-section for $\overline{K^0}$ is assumed to be three times larger than that of the K^0, the attenuation of the $\overline{K}^0$ beam is three times larger: 35 out of 50 will survive. When leaving the target, at the time t_c, the beam is made of $45/(35+45) = 56\%$ of K^0 and of 44% of $\overline{K}^0$. In terms of the K_1 and K_2 components, one has:

$$K_1(t_c) = \left(K^0(t_c) + \overline{K}^0(t_c)\right)/\sqrt{2} = (56+44)/\sqrt{2} = 70.7$$

$$K_2(t_c) = \left(K^0(t_c) - \overline{K}^0(t_c)\right)/\sqrt{2} = (56-44)/\sqrt{2} = 8.5$$

The relative ratios between the two components are:

$$K_1(t_c)/\left(K_1(t_c) + K_2(t_c)\right) = 89.3\%; \qquad K_2(t_c)/\left(K_1(t_c) + K_2(t_c)\right) = 10.7\%$$

The target thus becomes a K_1 regenerator.

Problem 12.4 If $\eta_{+-} = \varepsilon + \varepsilon'$ and $\eta_{00} = \varepsilon - 2\varepsilon'$, then:

$$|\eta_{00}|^2 = \varepsilon^2\left(1 - 2\frac{\varepsilon'}{\varepsilon}\right)^2$$

$$|\eta_{+-}|^2 = \varepsilon^2\left(1 + \frac{\varepsilon'}{\varepsilon}\right)^2$$

Using the approximation $\frac{1}{(1+x)^n} \simeq 1 - nx$ for $|x| < 1$ and neglecting second order terms in ε'^2, one has:

$$\frac{|\eta_{00}|^2}{|\eta_{+-}|^2} = \frac{(1-2\frac{\varepsilon'}{\varepsilon})^2}{(1+\frac{\varepsilon'}{\varepsilon})^2} \simeq \left(1 - 4\frac{\varepsilon'}{\varepsilon}\right)\left(1 - 2\frac{\varepsilon'}{\varepsilon}\right) \simeq \left(1 - 6\frac{\varepsilon'}{\varepsilon}\right).$$

Problem 12.5 From Eq. (12.15), the K_L is a superposition of $K^0, \overline{K}^0$ states:

$$|K_L\rangle = \frac{1}{\sqrt{2(1+|\varepsilon|^2)}}\left[(1+\varepsilon)\left|K^0\right\rangle - (1-\varepsilon)\left|\overline{K}^0\right\rangle\right] \tag{12.16}$$

The $\overline{K}^0$ can only decay with the emission of a negatively charged lepton, $\overline{K}^0 \to \pi^+\ell^-\overline{\nu}_\ell$. In terms of quark constituents, one has:

$$\overline{K}^0 = [\overline{d}s] \to (\overline{d}u)\ell^-\overline{\nu}_\ell = \pi^+\ell^-\overline{\nu}_\ell \tag{12.17}$$

where $s \to u\ell^-\overline{\nu}_\ell$ indicates the (Cabibbo suppressed) weak interaction transition. It is easy to verify that the transition $\overline{K}^0 \to \pi^-\ell^+\nu_\ell$ is forbidden. In a similar way, the K^0 can only decay to:

$$K^0 = [d\overline{s}] \to (d\overline{u})\ell^+\nu_\ell = \pi^-\ell^+\nu_\ell \tag{12.18}$$

The probability that the initial $|K_L\rangle$ evolves and decays into $\langle K_L \to \pi^+\ell^-\overline{\nu}_\ell|$ is by definition

$$\Gamma(K_L \to \pi^+\ell^-\overline{\nu}_\ell) = \langle \pi^+\ell^-\overline{\nu}_\ell | K_L \rangle. \tag{12.19}$$

Neglecting the factor $1/\sqrt{2(1+|\varepsilon|^2)}$ in (12.16) (which disappears later in the ratio) and remembering that $\langle K^0|\overline{K}^0\rangle = 0$, one can write:

$$\begin{aligned}\langle K_L|K_L\rangle &= \left[(1+\varepsilon)\langle K^0| - (1-\varepsilon)\langle \overline{K}^0|\right]\left[(1+\varepsilon)|K^0\rangle - (1-\varepsilon)|\overline{K}^0\rangle\right] \\ &= (1+\varepsilon)^2\langle K^0|K^0\rangle + (1-\varepsilon)^2\langle \overline{K}^0|\overline{K}^0\rangle\end{aligned}$$

Considering now the decay into $\pi^+\ell^-\overline{\nu}_\ell$, one can write:

$$\begin{aligned}\langle \pi^+\ell^-\overline{\nu}_\ell|K_L\rangle &= (1+\varepsilon)^2\langle \pi^+\ell^-\overline{\nu}_\ell|K^0\rangle + (1-\varepsilon)^2\langle \pi^+\ell^-\overline{\nu}_\ell|\overline{K}^0\rangle \\ &= 0 + (1-\varepsilon)^2\mathcal{B}\end{aligned} \tag{12.20}$$

where $\mathcal{B}$ is the branching ratio for the decay $\overline{K}^0 \to \pi^+\ell^-\overline{\nu}_\ell$. Similarly, for the $\pi^-\ell^+\nu_\ell$ decay, one has:

$$\begin{aligned}\Gamma(K_L \to \pi^-\ell^+\nu_\ell) = \langle \pi^-\ell^+\nu_\ell|K_L\rangle &= (1+\varepsilon)^2\langle \pi^-\ell^+\nu_\ell|K^0\rangle + 0 \\ &= (1+\varepsilon)^2\mathcal{B}\end{aligned} \tag{12.21}$$

The CPT theorem insures that the $\mathcal{B}$ in (12.20) and (12.21) are equal.

The observable quantity is the asymmetry:

$$\begin{aligned}A_L &= \frac{\Gamma(K_L \to \pi^-\ell^+\nu_\ell) - \Gamma(K_L \to \pi^+\ell^-\overline{\nu}_\ell)}{\Gamma(K_L \to \pi^-\ell^+\nu_\ell) + \Gamma(K_L \to \pi^+\ell^-\overline{\nu}_\ell)} \\ &= \frac{(1+\varepsilon)^2 - (1-\varepsilon)^2}{(1+\varepsilon)^2 + (1-\varepsilon)^2} = 2\frac{\mathrm{Re}\,\varepsilon}{1+\varepsilon^2} \simeq 2\mathrm{Re}\,\varepsilon\end{aligned} \tag{12.22}$$

The experimental result is $\mathrm{Re}\,\varepsilon = A_L/2 = (1.66 \pm 0.03)\cdot 10^{-3}$. The value obtained from non-leptonic decay is $|\varepsilon| = (2.23 \pm 0.01)\cdot 10^{-3}$ [Eq. (12.24)], with the phase $\phi \simeq 43.5°$ [Eq. (12.23)]. From this measurement, one derives that $\mathrm{Re}\,\varepsilon = |\varepsilon|\cos 43.5° \simeq 1.62$, in agreement within errors with the value obtained from the asymmetry measurement.

Problem 12.6 The B^0 is made of a $\overline{b}d$ quark-antiquark pair. Both the d and $\overline{b}$ can decay through the weak interaction:

$$\overline{b} \to \overline{c}W^+ \xrightarrow{BR=10.6\%} \overline{c}(\mu^+\nu_\mu) \tag{12.23}$$

$$d \to uW^- \xrightarrow{BR=10.6\%} u(\mu^-\overline{\nu}_\mu) \tag{12.24}$$

The second arrow reports the BR for the $W^{\pm}$ decay into a muon and a muon (anti)neutrino. If the decays (12.23) and (12.24) occur with the same frequency, a positive or negative muon can be present in the final state of a B^0 decay with the same probability. The $W^{\pm}$ have the same branching ratio for the decay in the charged lepton with the same sign, e.g. $BR(W^+ \to \mu^+\nu_\mu) = BR(W^- \to \mu^-\overline{\nu}_\mu) = 10.6\%$.

The values of CKM matrix elements corresponding to the above quark transitions implies that reaction (12.23) is disfavored with respect to (12.24) (see Sect. 8.14.3 and Problem 8.14). Indeed, the squared ratio between the corresponding matrix elements is $(V_{cb}/V_{ud})^2 = (0.04/0.97)^2 \sim 0.2\%$. On the other hand, Eq. (12.24) is disfavored due to the small value of the available energy E_0 in the final state. For the $\overline{b} \to \overline{c}$ decay, $E_0^b \simeq m_b - m_c \simeq 4300 - 1550 = 2750$ MeV. For the $d \to u$ decay, $E_0^d \simeq m_d - m_u \simeq 2 \div 5$ MeV. As both reactions (12.23) and (12.24) are three-body decays, the Sargent rule (8.18) can be applied. Thus, the decay (12.24) is suppressed by a factor $\sim (10^3)^5 \sim 10^{15}$ with respect to (12.23). Taking into account both the CKM matrix element ratio and the phase space factor, the estimated ratio between the BR of the two considered decays is:

$$\frac{BR(\overline{b} \to \overline{c}W^+)}{BR(d \to uW^-)} \sim \left(\frac{E_0^b}{E_0^d}\right)^5 \left(\frac{V_{cb}}{V_{ud}}\right)^2 \simeq 10^{15} \cdot 10^{-3} = 10^{12}$$

It is therefore very unlikely to observe negative muons in B^0 decays.

Note that in the situation sketched in Fig. 12.8, a $u\overline{u}$ pair is created during the hadronization phase; in this case, the reaction is:

$$\overline{b}d \to (\overline{c}W^+)d \longrightarrow \overline{c}(u\overline{u})d + W^+ = (\overline{c}u)(\overline{u}d) + (\mu^+\nu_\mu) = \overline{D}^0\pi^-\mu^+\nu_\mu$$

Problem 12.7 The reaction rate, i.e., the number of interactions per time unit is

$$R = (\sigma \Phi N_{Cl})\ \mathrm{s}^{-1} \tag{12.25}$$

where σ (cm^2) is the reaction cross-section, N_{Cl} the number of ^{37}Cl nuclei in the detector and Φ (cm^{-2} s^{-1}) the flux of neutrinos with enough energy to induce the reaction. The total mass of the liquid in the detector is

$$m = \rho v = \left(1.5\ \mathrm{g\,cm^{-3}}\right)\left(4 \cdot 10^8\ \mathrm{cm^3}\right) = 6.0 \cdot 10^8\ \mathrm{g}$$

The number of moles present is $n = \frac{m}{P} = \frac{6 \cdot 10^8}{164} = 3.6 \cdot 10^6$ moles.
The number of C_2Cl_4 molecules is:

$$N_{mol} = nN_A = \left(3.6 \cdot 10^6\ \text{moles}\right)\left(6 \cdot 10^{23}\ \text{moles}^{-1}\right) \simeq 2.2 \cdot 10^{30}\ \text{molecules}$$

where N_A is Avogadro's number. There are 4 Cl atoms per molecule, and the total number of the particular isotope of the chlorine atoms (^{37}Cl) in the detector is:

$$N_{Cl} = 0.25 \times (4N_{mol}) = 2.2 \cdot 10^{30}$$

(the factor 0.25 takes into account the final isotopic abundance of ^{37}Cl).

From the solar luminosity and taking into account the Earth-Sun distance $D = 150 \cdot 10^9$ m, the energy flux S can be derived:

$$S = \frac{L_\odot}{4\pi D^2} = \frac{3.84 \cdot 10^{26}}{4\pi(1.5 \cdot 10^{11})^2} = 1358\ \mathrm{W\,m^{-2}} = 8.5 \cdot 10^{11}\ \mathrm{MeV\,s^{-1}\,cm^{-2}}$$

where the conversion factor 1 J $= 0.63 \cdot 10^{13}$ MeV is used in the last step.

It follows that the flux of neutrinos arriving on Earth (7% of the Sun thermonuclear energy), is:

$$\Phi_\nu = 0.07\frac{S}{\langle E\rangle} = 0.07\frac{8.5 \cdot 10^{11}\ \mathrm{MeV\,cm^{-2}\,s^{-1}}}{1\ \mathrm{MeV}} = 6 \cdot 10^{10}\ \mathrm{cm^{-2}\,s^{-1}}. \qquad (12.26)$$

Since only 0.1% of neutrinos have an energy sufficient to give rise to the reaction, the corresponding neutrino flux is:

$$\Phi = \Phi_\nu/1000 = 6 \cdot 10^7\ \mathrm{cm^{-2}\,s^{-1}}$$

Inserting in Eq. (12.25) the values of Φ, N_{Cl} and remembering that $\sigma = 10^{-43}$ cm^2, one finds:

$$\begin{aligned} R = \sigma\Phi N_{Cl} &= \left(10^{-43}\ \mathrm{cm^2}\right)\left(6 \cdot 10^7\ \mathrm{cm^{-2}\,s^{-1}}\right)\left(2.2 \cdot 10^{30}\right) \\ &= 1.3 \cdot 10^{-5}\ \mathrm{events\,s^{-1}} = 1.3 \cdot 10^{-5} \cdot 3600 \cdot 24 \\ &= 1.1\ \mathrm{events\,day^{-1}} \qquad (12.27) \end{aligned}$$

Traditionally, the number of events due to solar neutrinos is expressed in SNU (1 SNU = 1 count/s for 10^{36} target atoms). The theoretical prediction for the experiment in question is therefore about 6 SNU: this value is obtained using $N = 10^{36}$ in Eq. (12.27) [$R = (10^{-43}\ \mathrm{cm^2})(6 \cdot 10^7\ \mathrm{cm^{-2}\,s^{-1}})(10^{36})$]. The value obtained with a more detailed calculation is 8 SNU. The value measured by the *Clorine* (or Davis) experiment, averaged over a period of more than 20 years, was 2.6 SNU. This has given rise to the so-called *solar neutrino problem*. Raymond Davis Jr. was awarded the Nobel Prize for the first experimental measurement of neutrinos from the Sun in 2002. The solar neutrino problem was solved theoretically by assuming that neutrinos have a small mass and can *oscillate*. Neutrino oscillations were experimentally confirmed by many detectors using different experimental techniques.

Problem 12.9

(a) See Fig. 12.4.
(b) The function factor, $\sin^2(\pi L/L_{osc})$, of the oscillation probability given in (12.49), passes through a maximum when $\pi L/L_{osc} = N\pi/2$ where N is an integer. Making use of Eq. (12.50), this corresponds to the condition $L\Delta m^2/2.48E_\nu = N/2 \rightarrow L\Delta m^2/1.24E_\nu = N$. In the energy range from 5 <

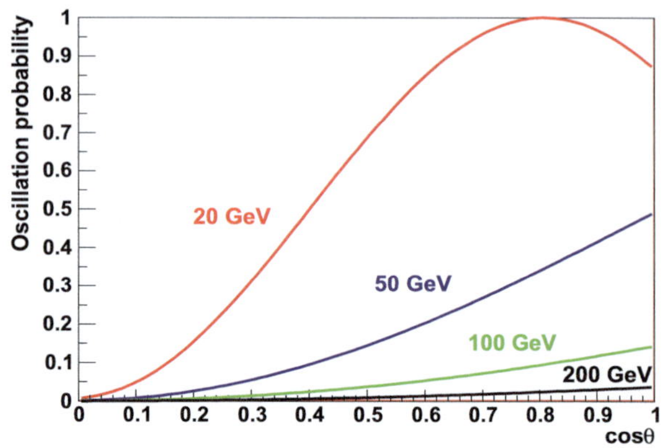

Fig. 12.4 Oscillation probability as a function of the zenith angle cosine for four different energies. $\cos\theta = 1$ represents a vertical upgoing neutrino. The formula (12.49) was used, with L given by (12.7). The ROOT package [10w3] was used to draw the plot

$E_\nu < 200$ GeV, the left member of the latter equation ranges from about 0.1 to 5 (using $L = 2R_T$ and $\Delta m^2 = 2.4 \cdot 10^{-3}$ eV^2). This means that, in the considered energy range, the function $\sin^2(\pi L/L_{osc})$ passes through 5 maxima, i.e., for $N = 1, 2, 3, 4, 5$. The observed value corresponds to the average of the squared sinus function, i.e., $\overline{\sin^2(\pi L/L_{osc})} = 0.5$.

The expected flux of muon neutrinos from the nadir direction is reduced by 50% in the energy region where MACRO observed neutrino-induced muons. The same situation occurs for SuperKamiokande (see Fig. 12.14).

Problem 12.10 Let consider the probability that a muon produced by a pion decay, decays in turn to an electron (the same for the $K \to \mu \to e$), as discussed in Problem 10.10. In the case of atmospheric neutrinos, the path length L corresponds to a large fraction of atmospheric depth. For the three given muon energies ($E_\mu = 0.2$, 20 and 200 GeV), we have from (10.10):

$$P(\mu \to e) \simeq 100\%; \quad 15\%; \quad 1.5\%$$

respectively. This corresponds to the fact that at low energies (sub-GeV) all muons decay and the ratio

$$R = \frac{(\nu_e + \overline{\nu}_e)}{(\nu_\mu + \overline{\nu}_\mu)} = \frac{1}{1+1} = 0.5 \tag{12.28}$$

as for each decayed π^+ there is a ν_μ and a μ^+, which in turns decays into positron, $\overline{\nu}_\mu$, ν_e (the same, with charge conjugated particles, for the π^- decay). At high energies, the ratio tends to zero, as the muon has not enough *space* to decay before arriving on Earth.

Problem 12.12

(a) Using the mixing matrix (12.51) and the approximation for a neutrino with dominant mass (see Sect. 12.6.3), one has:

$$P(\nu_\alpha \to \nu_\beta) = 4|U_{\alpha 3}|^2|U_{\beta 3}|^2 \sin^2\left(\frac{\Delta m^2_{13}}{4E}L\right)$$
$$= 4|U_{\alpha 3}|^2|U_{\beta 3}|^2 \sin^2 \Phi_{23}$$

where α, β indicate two different neutrino flavors (e, μ, τ) and

$$\Phi_{23} = \frac{\Delta m^2_{23}}{4E_\nu}L \simeq \frac{\Delta m^2_{13}}{4E_\nu}L.$$

For instance, for $\nu_\mu \to \nu_e$, the corresponding oscillation probability (12.9a) is obtained with $\alpha = \mu$ and $\beta = e$:

$$P(\nu_\mu \to \nu_e) = 4|U_{\mu 3}|^2|U_{e3}|^2 \sin^2 \Phi_{23}$$
$$= 4 \sin^2\theta_{23} \cos^2\theta_{13} \sin^2\theta_{13} \sin^2\Phi_{23}$$
$$= \sin^2\theta_{23} \sin^2 2\theta_{13} \sin^2\Phi_{23}$$

where the last member is obtained applying the trigonometric relation $\sin 2\theta = 2\sin\theta\cos\theta$.

(b) There are two possibilities:
1. Using the disappearance technique of $\overline{\nu}_e$ from reactors, through comparisons of expected and measured neutrino flux with the formula (12.9e). For CPT, the same oscillation formulas hold for ν and $\overline{\nu}$.
2. Using the appearance technique of ν_e from a ν_μ beam. In this case, formula (12.9a) is used. Remember that the quantity $\sin^2 2\theta_{23} \simeq 1$.

A reactor neutrino oscillation experiment at the Chooz nuclear power station in France [12C03] was the first $\overline{\nu}_e$ disappearance experiment (immediately followed by Palo Verde). The detector was located in an underground laboratory with 300 m.w.e. (meter water equivalent) overburden rock, at $\sim$1 km from the neutrino source. It consisted of a 5-ton target filled with 0.09% gadolinium loaded liquid scintillator, surrounded by an intermediate 17-ton and outer 90-ton regions filled with undoped liquid scintillator. Reactor $\overline{\nu}_e$ were detected via the usual inverse β decay reaction (8.19). Gd-doping was chosen to maximize the neutron capture efficiency. The CHOOZ experiment found no evidence for ν_e disappearance, with a 90% C.L. upper limit for $\sin^2 2\theta_{13} < 0.19$. The Double Chooz, Reno and Daya Bay experiments plan to measure this mixing angle, or to improve the upper limit, with similar techniques.[1]

[1]During the proof correction of this book, the Daya Bay Reactor Neutrino Experiment has measured a non-zero value for the neutrino mixing angle θ_{13} with a significance of 5.2 standard de-

In the accelerator neutrino oscillation experiments with conventional *off-axis* neutrino beams (see Problem 10.11), the small θ_{13} can be measured using the appearance of electron neutrino in a muon neutrino. The *Tokai to Kamioka* (T2K) experiment is a long baseline neutrino experiment designed to measure the mixing angle in this way. A strong, artificial neutrino beam, produced in J-PARC accelerator complex in Tokai is directed to the Kamioka mine 295 km away from the beam source. Neutrinos undergo the oscillation process and reach Super-Kamiokande, a large water Cherenkov detector situated in the mine. The J-PARC complex suffered from the March 11, 2011 earthquake and tsunami in Japan. In the data collected until then, 6 ν_e candidates were found with an expected number of background events [12A11] of 1.5. This is a 2.5σ observation which the collaboration aims to improve in the next years.

References

[12K89] Kaiser, B.: The Physics of Massive Neutrinos. World Scientific, Singapore (1992). ISBN: 9971-50-661-0

[12A98] Ambrosio, M., et al. (MACRO Collaboration): Measurement of the atmospheric neutrino-induced upgoing muon flux using MACRO. Phys. Lett. **434**, 451 (1998)

[12K00] Klapdor-Kleingrothaus, H.V., Pas, H., Smirnov, A.Y.: Neutrino mass spectrum and neutrinoless double beta decay. Phys. Rev. D **63**, 073005 (2001). hep-ph/0003219

[12C03] Apollonio, M., et al.: Search for neutrino oscillations on a long base-line at the CHOOZ nuclear power station. Eur. Phys. J. C **27**, 331 (2003)

[12K04] Kajita, T.: Atmospheric neutrinos. New J. Phys. **6**, 194 (2004). http://iopscience.iop.org/1367-2630/6/1/194

[12A08] Avignone, F.T., Elliott, S.R., Engel, J.: Double beta decay, Majorana neutrinos, and neutrino mass. Rev. Mod. Phys. **80**, 481–516 (2008). arXiv:0708.1033 [nucl-ex]

[12A11] Abe, K., et al. (T2K Collaboration): Indication of electron neutrino appearance from an accelerator-produced off-axis muon neutrino beam. Phys. Rev. Lett. **107**, 041801 (2011)

[12C11] Andreotti, E., et al. (Cuoricino Collaboration): ^{130}Te neutrinoless double-beta decay with CUORICINO. Astropart. Phys. **34**, 822–831 (2011)

viations. Antineutrinos from six 2.9 $\mathrm{GW_{th}}$ reactors were detected in six antineutrino detectors deployed in two near (distance = 470 m and 576 m) and one far (1648 m) underground experimental halls. With 55 days of data, more than 10000 $\overline{\nu}_e$ interactions were detected at the far hall and ~80000 in the near halls. While in the near hall this number is in agreement with expectation, the ratio of the observed to expected number of $\overline{\nu}_e$ at the far hall is $R = 0.940 \pm 0.011(\mathrm{stat}) \pm 0.004(\mathrm{syst})$. This result is explained assuming $\sin^2 2\theta_{13} = 0.092 \pm 0.016(\mathrm{stat}) \pm 0.005(\mathrm{syst})$. Details on arXiv:1203.1669v1 [hep-ex].

Chapter 13
Microcosm and Macrocosm

Problems

13.1. **Neutrinos from neutron decay**. It is believed that cosmic rays are accelerated through an iterative mechanism (whose theoretical model is due to E. Fermi), which consists of a sequence of collisions of charged particles with the shock wave produced by the explosion of a supernova. In each collision, a particle gains a small amount of energy. Due to the deflection caused by magnetic fields, charged particles have a low probability to escape the acceleration region. The situation is different for neutral particles such as neutrons; they are not subjected to magnetic fields and can elude the acceleration region.
(a) Give a process in which neutrons can be produced.
(b) What is the minimum energy necessary for a neutron to escape the acceleration region with a probability $\geq 1/e$ ($\tau_n = 887$ s)? Assume that the region size is the order of one light year.
(c) What would be the maximum value of the angle formed by the final state electron (or neutrino) with the flight direction of the neutron?
[See solutions]

13.2. **Search for the proton decay**. The Kamiokande detector had a fiducial volume of 1000 tons of water. Calculate the number of protons in the detector fiducial volume. If the proton has a lifetime of 10^{32} years, how many protons would decay in 1000 tons of water each year?
[See solutions]

13.3. **Indirect estimate of the proton lifetime**. Geophysical measurements show that the Earth emits approximately 40 TeraWatt of energy (see also Problem 14.14). Assuming that all this energy is due to the proton decay process, whose rest mass transforms into heat, estimate the proton lifetime.
[See solutions]

S. Braibant et al., *Particles and Fundamental Interactions: Supplements, Problems and Solutions*, Undergraduate Lecture Notes in Physics,
DOI 10.1007/978-94-007-4135-5_13,

13.4. **Search for massive exotic particles in Cosmic Rays**. New exotic heavy particles X (e.g., with mass $m_X = 100$ GeV) can be searched studying the arrival time of hadrons in showers produced in high energy cosmic ray collisions with atmospheric nuclei (see Supplement 1.1). Assuming that these particles travel 2 km before decay, calculate:

(a) the threshold energy to produce a pair of X particles;
(b) the energy of a particle in the laboratory system for a X produced at rest in the c.m. system;
(c) the time delay of the massive X particle with respect to the lighter hadrons in the showers, moving at the light speed c;
(d) the time delay if the particle has mass $m_X = 1$ TeV.

[See solutions]

13.5. **Gravitational binding energy**.

(a) Show that the gravitational potential energy of a spherical mass M with uniform density and radius R is $V_G = -3G_N M^2/5R$.
(b) Calculate the gravitational potential energy of a mass of material equivalent to one solar mass with uniform density and with a radius R equal to
  1. 1 light-year (*protostar*)
  2. 1 solar radius (*star*)
  3. 1000 km (*white dwarf*)
  4. 10 km (*neutron star*).

[See solutions]

13.6. **Supernovae and neutrinos**. The 1987A supernova (SN) was located in the Large Magellanic Cloud at around 170000 light-years from Earth. During a SN event, a neutrino burst lasting ~ 2 s is expected. About 10 neutrino interactions from the SN1987A in 1000 t of water were observed in a time interval Δt of less than 10 seconds (assume the value $\Delta t = 3$ s in the calculation). The average energy of supernova neutrinos is about 12 MeV (the energy range is between $E_{min} = 5$ MeV and $E_{max} = 20$ MeV). If massive, the neutrinos with higher energy are expected to arrive on Earth before the lower energy neutrinos. Estimate:

(a) an upper limit for the neutrino mass from the measured neutrino arrival time interval Δt;
(b) the total energy released in neutrinos by the SN;
(c) the gravitational binding energy $3G_N M^2/5R$ of the star, taking into account that the neutron star formed in the explosion has mass $1.4M_\odot$ and radius $R = 10$ km.

[See solutions]

13.7. **Propagation of protons in the cosmic microwave background radiation: the GZK effect**. The Universe is filled with a cosmic microwave background radiation (CMBR) with a black body spectrum and an average temperature of ~2.7 K ($\overline{E} = 10^{-3}$ eV). The average density of CMBR photons is $\rho_\gamma = 400\ \text{cm}^{-3}$. The high energy cosmic ray protons, originated from extragalactic sources, can interact with the microwave cosmic background radiation through the resonant reaction:

$$\begin{aligned} p^+\gamma \to \Delta^+ &\to \pi^+ n \\ &\to \pi^0 p \end{aligned}$$

(a) Find the threshold energy that protons must have to induce this reaction.
(b) Calculate the average distance traveled by a proton before undergoing the resonant reaction, knowing that the cross-section of the process is $\sigma_{p\gamma} = 250\ \mu\text{b}$.

This process limits the size of the Universe from which ultra-high energy protons can arrive. This is the so-called Greisen-Zatsepin-Kuzmin (GZK) cut-off named after the physicists that first hypothesized it.
[See solutions]

13.8. **γ-rays attenuation in the CMBR**. High energy gamma-rays can interact with lower-energy photons through the process $\gamma + \gamma \to e^+ + e^-$. This process has a cross-section $\sigma = (8\pi/9)r_e^2$ where $r_e = 2.8 \times 10^{-15}$ m is the classical electron radius.

(a) Find the threshold energy of the high energy gamma-rays so that the reaction can take place through the interaction with
   1. the cosmic microwave background radiation;
   2. infrared photons (~ 0.1 eV);
   3. optical photons (~ 2 eV).
(b) For these three cases, calculate the average distance travelled by the high energy gammas before being converted. Compare the results with the size of the Universe.

[Hint: use the solution of Problem 13.7]

13.9. **Neutrino telescopes**. A neutrino telescope (NT) detects secondary particles produced in neutrino interactions as $\nu_\mu N \to \mu X$ (see Supplement 13.1). The neutrino-induced muons travel in a volume of 1 km^3 of ice or water, where a number N_{pmt} of photomultipliers are plunged. Assume that:

1. The muon track length is 1 km;
2. A muon emits 350 Cherenkov photons per cm of water (Sect. 2.11);
3. The PMT have a 10″ diameter (1 inch = 2.54 cm) and a quantum efficiency $\epsilon_{pmt} \simeq 0.25$;
4. The PMTs are inside optical module, with light collection transparency of 80%;
5. The water absorption length is $\lambda_{abs} = 50$ m in the 400–500 nm range (100 m for the ice);

6. The number of detected photoelectrons necessary to reconstruct a muon track is $N_{p.e.} = 100$.

Estimate the number of optical sensor N_{pmt} needed to track a muon.

This number is one of the major impact factors on the cost of an experiment.

[See solutions]

13.10. **Cosmic Rays in galactic magnetic field**. The Galaxy is filled with magnetic fields with average modulus of $B = 3\ \mu\text{G}$ and directions coherent for a length of $1 \div 10$ pc. Evaluate the curvature radius of a cosmic ray proton with 10^{12}, 10^{15} and 10^{18} eV in the galactic magnetic field. Compare the curvature radius with the Galaxy dimensions (radius $\sim$15 kpc, thickness $\sim$200 pc). [Hint: see Problem 3.2. A: 3 10^{-4}, 0.3, 300 pc.]

13.11. **Cosmic accelerators of Cosmic Rays**. Using dimensional arguments, estimate the maximum energy of a charged particle (with charge Ze) accelerated in a strongly varying magnetic field. Apply the relation to the case of a neutron star rapidly rotating around its axis (*pulsar*) with mass $M = 1.4M_{\odot}$ and magnetic field $B = 10^7$ T. Explain how such magnetic field can be generated and estimate the pulsar angular velocity.

[See solutions]

13.12. **Search for magnetic monopoles.** A muon with energy $E = 10$ GeV and a (hypothetical) magnetic monopole (MM) with magnetic charge $g = 68.5e = g_D$ and speed $v_1 = 0.01c$ crosses a 25 cm thick layer of liquid scintillator (for instance that of the MACRO experiment at Gran Sasso laboratory, with density $\rho = 0.85\ \text{g cm}^{-3}$). Evaluate for the muon and for the (hypothetical) MM:

(a) The total energy lost in the liquid scintillator;
(b) The energy lost in the scintillator that produces light;
(c) The total energy lost for a MM with $v_2 = 0.3c$. Show that MM with $\beta > 0.1$ behave as particles with an equivalent electric charge $Ze = g\beta$.

Refer to the energy loss given in Fig. 2.2(b) for charged particles and that given in Fig. 13.1 for particles with magnetic charge.

[See solutions]

Supplement 13.1: Cosmic Accelerators

Some galactic accelerators must exist to explain the presence of Cosmic Rays (CRs) with energies up to the 10^{19} eV (see Supplement 1.1 and Problem 13.11). It is roughly expected that galactic accelerators are related to the final stage of the evolution of massive, bright and relatively short-lived stellar progenitors [L92].

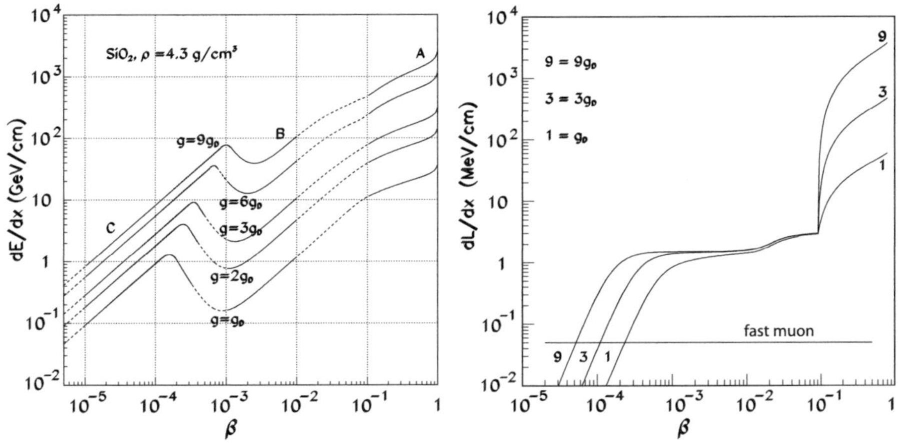

Fig. 13.1 (**a**) Energy loss of magnetic monopoles with magnetic charge $g = 1, 2, 3, 6, 9g_D$ as a function of $\beta = v/c$ in Si (density $\rho = 4.3$ g cm^{-3}; this is a relatively high density, similar to that present inside the Earth). (**b**) Energy loss in form of visible light dL/dx (in MeV/cm) produced by a magnetic monopole with magnetic charge $g = g_D, 3g_D, 9g_D$ and for a relativistic muon in a scintillator as a function of β

Due to the influence of galactic magnetic fields (Problem 13.10), charged CRs do not point to the sources. Neutral particles (gamma-rays and neutrinos) do not suffer the effect of magnetic fields: they are among the decay products of accelerated charged particles but cannot be directly accelerated.

The astrophysical production of high energy gamma-rays and neutrinos is supposed to occur mainly through the decay of neutral and charged pions produced in interactions of high energy protons with the matter or photon fields surrounding the source. Accelerated protons interact with photons in the surroundings of the CRs emitter, predominantly via the Δ^+ resonance:

$$\begin{aligned} p+\gamma &\to \Delta^+ \to \pi^o + p \\ p+\gamma &\to \Delta^+ \to \pi^+ + n. \end{aligned} \tag{13.1}$$

Protons also interact with the ambient matter (protons, neutrons and nuclei), giving rise to the production of charged and neutral mesons:

$$p + N \to \pi^{\pm}, \pi^o + X. \tag{13.2}$$

Very high energy γ-rays ($E_\gamma > 100$ MeV) and neutrinos are produced by the meson decays. Neutral mesons decay in photons (observed on Earth as γ-rays):

$$\pi^o \to \gamma\gamma \tag{13.3}$$

while charged mesons decay into final states with neutrinos:

$$\begin{aligned} &\pi^+ \to \nu_\mu + \mu^+ \\ &\qquad\qquad \hookrightarrow \mu^+ \to \overline{\nu}_\mu + \nu_e + e^+ \\ &\pi^- \to \overline{\nu}_\mu + \mu^- \\ &\qquad\qquad \hookrightarrow \mu^- \to \nu_\mu + \overline{\nu}_e + e^- \end{aligned} \tag{13.4}$$

Cosmic Rays, Gamma-Rays and Neutrinos The accelerator mechanisms that produce CRs, can also produce high energy photons and neutrinos through the processes (13.3) and (13.4). In this scenario (*hadronic model*) the energy escaping from the source is distributed between CRs, γ-rays and neutrinos.

The energy spectrum of CRs is a power law $E^{-\alpha_S}$, where the spectral index $\alpha_S \simeq 2$ (see Supplement 1.1). The multiplicity of charged particles produced by hadronic interactions like (13.2) has the smooth $\ln s$ dependence (see Fig. 9.19). Only a small fraction of the c.m. energy is converted into mass. Therefore, due to energy conservation, most of the initial proton energy transforms in kinetic energy of the outgoing particles. The produced charged mesons have the same energy spectrum as the initial CR proton, as well as their daughter γ-rays and neutrinos from (13.3, 13.4). For this reason, it is expected a power low spectrum also for γ and ν, with $\alpha_S \sim \alpha_\nu \sim \alpha_\gamma$.

Electrons accelerated by astrophysical objects can produce γ-rays but not neutrinos (*leptonic model*). The most important process which produces high energy γ-rays in the leptonic model is the Inverse Compton (IC) scattering.[1] IC γ-rays are produced in the interactions of energetic electrons with ambient background photon fields: the cosmic microwave background radiation, and the diffuse galactic radiation of star light. Most of observed TeV γ-ray galactic sources (see below) have a power law energy spectrum $E^{-\alpha_\gamma}$, where $\alpha_\gamma \sim 2.0 \div 2.5$. The values of the spectral index are very close to the expected spectral index of CR sources, α_S. This leads to the conclusion that some of the TeV γ-rays sources can also be the sources of galactic CRs.

MeV-GeV Photons Sources emitting MeV-GeV photons were first observed in the 1990's by the Energetic Gamma-ray Experiment Telescope (EGRET) on board of the CGRO satellite. Following its launch in June 2008, the Fermi Gamma-ray Space Telescope (Fermi) began a sky survey which produced a deeper and better resolved map of the γ-ray sky than any previous space mission. The initial result for energies above 100 MeV [13F09] regards the 205 most significant γ-ray sources, which are the best characterized and best localized ones. Most of them are in the galactic plane, and were associated with known pulsars (see Supplement 14.2).

[1]The inelastic scattering of photons in matter results in a decrease in energy of an X-ray or gamma-ray photon; this is the so-called Compton effect. The Inverse Compton effect corresponds to the fact that a low energy photon is scattered to higher energies by a high energy electron.

Photons with $E > 100$ GeV Gamma-rays above 100 GeV are detected on ground, using the Imaging Air-Cherenkov Technique (IACT) [13A08, 13D08]. When impinging the Earth's atmosphere, the high energy γ-rays collide with the nuclei present in the atmosphere, producing a cascade (shower) of high energy relativistic secondary particles. These emit Cherenkov photons, at a characteristic angle in the visible and UV range, which move through the atmosphere. The atmospheric showers induced by cosmic rays can therefore be observed on the surface of the Earth through the detection of the Cherenkov light.

The pioneering ground based γ-ray experiment was built by the Whipple collaboration in the 1990's. During the last decade, several ground based γ-ray detectors were developed, both in North and South Earth hemispheres. At present, the new generation apparatuses are the H.E.S.S., VERITAS and MAGIC telescopes. These IACT telescopes have produced a catalogue of TeV γ-ray sources, with a concentration of sources in the galactic plane.

Possible Galactic and Extragalactic Sources It is commonly believed that about 10% of the energy emitted in galactic supernovae (SN) explosions at an approximate rate of 3 per century can provide the power needed to account for the observed cosmic rays up to $\sim 10^{15}$ eV through the iterative Fermi mechanism [1F49a]. If the final product of the SN is a neutron star, already accelerated particles can gain additional energy, due to the neutron star strong magnetic field. Shell-type **SN Remnants** (SNR) are considered to be the most likely sites of galactic CR acceleration, hypothesis supported by recent observations from the TeV γ-ray IACT.

Figure 13.2 [13O02] shows a diagram first produced by Hillas (1984). Hillas derived the maximum energy which a particle of a given charge Ze can reach, independently of the acceleration mechanism (see also Problem 13.11). It was obtained from the simple argument that the particle radius in the magnetic field $B_{\mu G}$ (in units of 10^{-6} Gauss) should be smaller than the size R_{kpc} (in kpc) of the acceleration region. This energy $E(\text{EeV})$ (in units of 10^{18} eV) is given by:

$$E(\text{EeV}) \sim \beta Z B_{\mu G} R_{kpc} \tag{13.5}$$

where β is the velocity of the shock wave in the Fermi model or in any other acceleration mechanism. Figure 13.2 gives the relation between the dimensions of the astrophysical objects and the magnetic fields needed to contain the accelerating particle.

Other galactic candidates as CR sources are the **pulsar wind nebulae**. They differ from the shell-type SNRs because there is a pulsar in the centre which blows out equatorial winds and, in some cases, jets of very fast moving material into the nebula. **Microquasars** are galactic X-ray binary systems, which exhibit relativistic radio jets, observed in the radio band. The name is due to the fact that they are morphologically similar to the so-called quasars, i.e., very energetic and distant Active Galactic Nuclei (AGN). Indeed, the presence of jets makes the microquasars similar to small quasars. This resemblance could be more than morphological: the physical processes that govern the formation of the accretion disk and the plasma ejection in microquasars are probably the same ones as in large AGN.

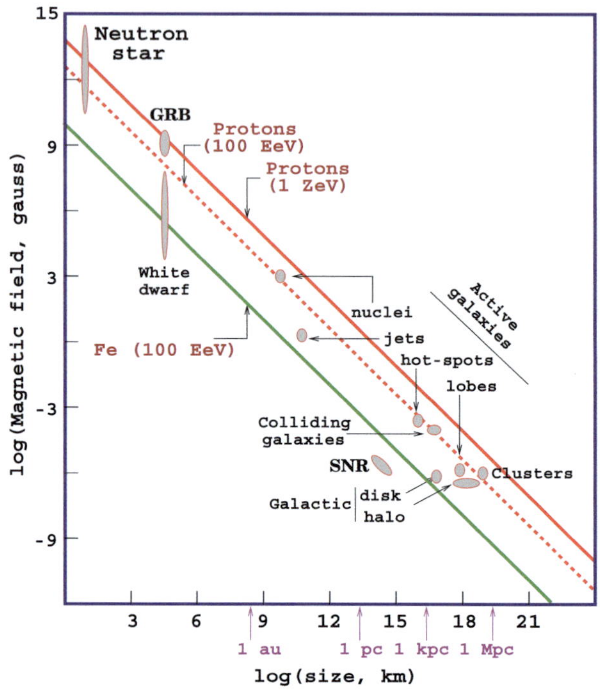

Fig. 13.2 Example of the diagram first produced by Hillas. Acceleration of cosmic rays up to a given energy requires magnetic fields and sizes above the respective line. Some source candidates are still controversial (1 EeV = 10^{18} eV, 1 ZeV = 10^{21} eV)

Ultra High Energy (UHE) Cosmic Rays ($>10^{19}$ eV) are likely of extragalactic origin. **Active Galactic Nuclei** (AGN) or **gamma-ray bursts** (GRBs) are the principal candidates. AGN are galaxies with a very bright core of emission embedded in their centre, where a supermassive black hole ($10^6 \div 10^9$ solar masses) is probably present. GRBs are short flashes of γ-rays, lasting typically from milliseconds to tens of seconds, and carrying most of their energy in photons of MeV scale. A possible origin of the GRBs with duration of tens of seconds is the collapse of massive stars into black holes.

Detectors for Cosmic Neutrinos A large volume neutrino detector behaves as a *telescope* when the neutrino direction is reconstructed with an angular precision of 1° or better. This is the case for high energy charged current ν_μ interactions [13C10].

The small interaction cross-section of neutrinos allows them to come from far away, but it also consists in a drawback, as their detection requires a large target mass. The idea of a neutrino telescope based on the detection of the secondary particles produced in neutrino interactions was first formulated at the beginning of the 1960's by Markov. He proposed *to install detectors deep in a lake or in the sea and to determine the direction of the charged particles with the help of Cherenkov radiation*. Starting from the Markov idea and from the present knowledge of TeV γ-ray sources, the challenge to detect cosmic neutrinos is open for a multi-km^3 scale apparatus with at least 5000 detector units (see Problem 13.9).

The pioneering project for the construction of an underwater neutrino telescope was due to the DUMAND collaboration, which attempted to deploy a detector off the coast of Hawaii in the 1980s. At that time, the technology was not enough advanced to overcome the challenge and the project was cancelled. In parallel, the BAIKAL collaboration started to work in order to realize a workable detector system under the surface of the Baikal lake.

A major step towards the construction of a large neutrino detector in deep ice was due to the AMANDA collaboration. AMANDA operated optical sensors placed in the ice layer of the Antarctic starting from 1993. After the detector completion in 2000, the AMANDA collaboration proceeded with the construction of a much larger apparatus, the IceCube detector [13w3], whose construction was completed early 2011.

In water, the pioneering DUMAND experience is being continued in the Mediterranean Sea by the ANTARES experiment. The final goal is the construction of a km^3-scale detector in the Mediterranean Sea. KM3NeT [13w4] is an European deep-sea research infrastructure, which will host a neutrino telescope with a volume of at least one cubic kilometer at the bottom of the Mediterranean Sea.

Fiat lux. It was written, and scientists never fail to observe new spectacular astrophysical discoveries when new experimental techniques on new photons wavelengths become available: from the Cosmic Microwave Background observation up to the TeV γ-ray astronomy using Imaging Air-Cherenkov Technique.

Fiat neutrinos, it was never written, and Mr. Pauli himself has feared that this particle would never be discovered. Nevertheless, observation of solar neutrinos and of neutrinos from the supernova 1987A opened up a new observation field. High energy neutrino astronomy is a young discipline derived from the fundamental necessity of extending conventional astronomy beyond the usual electromagnetic messengers.

Solutions

Problem 13.1

(a) Neutrons can be produced through the interaction of protons with the electromagnetic radiation in the environment surrounding the supernova through the resonant formation of a Δ baryon:

$$p + \gamma \to \Delta^{+} \to \pi^{+} + n.$$

In addition, neutrons can be produced in the interaction of high energy protons with other protons in the supernova remnant environment.

(b) A high energy neutron ($\beta \simeq 1$) in the observer's system has a decay length $L = \Gamma \tau_n c$, where $\Gamma = E/(m_n c^2)$ (see Sect. 3.6.1) is the Lorentz factor and τ_n, the neutron lifetime in the rest frame system.

The survival probability is $P = e^{-L/c\Gamma\tau_n}$, and $P = 1/e$ when $L = c\Gamma\tau_n$. The length given in the problem is one light-year. Therefore, one has (1 year = $3.15 \cdot 10^7$ s):

$$L = 3.15 \cdot 10^7 c = \Gamma\tau_n c \longrightarrow \Gamma = \frac{3.15 \cdot 10^7}{887} = 3.5 \cdot 10^4$$

which corresponds to a neutron energy of:

$$E = \Gamma m_n c^2 \simeq 33000 \text{ GeV} = 33 \text{ TeV}.$$

Any larger value of Γ gives a higher survival probability.

(c) The maximum kinetic energy for the electron (or neutrino) in the beta decay is $T_\beta = E_\beta - m_e \simeq 0.78$ MeV (see Sect. 8.3.1). Assuming that the electron (or neutrino) is emitted in the direction perpendicular to the direction of motion of the neutron with energy E, the angle in the observer system is:

$$\tan\theta \simeq \theta = T_\beta/E \simeq 2.3 \cdot 10^{-8}.$$

While the electron can be deflected in its motion by magnetic fields, neutrinos "remember" the original direction of the neutron. The detection of neutrinos from astrophysical sources could help to identify the sources (still unknown) of cosmic rays detected on Earth (see Supplement 1.1).

Problem 13.2 The number of atoms in one cm^3 of a material with mass number A is given by $N_A\rho/A$, where N_A is Avogadro's number. For a molecule of water, the mass number is $A = 18$, and the number of protons per molecule is $Z = 8 + 1 + 1$. Since 1000 t of water occupy a volume of $V = 10^9$ cm^3, the total number of protons in the fiducial volume of the Kamiokande detector was:

$$N = \frac{N_A \rho V Z}{A} = \frac{6 \cdot 10^{23} \times 10^9 \times 10}{18} = 3.3 \cdot 10^{32}. \tag{13.6}$$

Denoting with τ_p the proton lifetime, the number of surviving protons, during an observation time t, is $N(t) = Ne^{-t/\tau_p}$. Since $t \ll \tau_p$, the series expansion at the first order can be used and one has: $N(t) = Ne^{-t/\tau_p} = N(1 - \frac{t}{\tau_p})$. If the proton lifetime is 10^{32} years, the number of protons decaying in one year is:

$$N_D = N - N(t) = N\frac{t}{\tau_p} = 3.3 \cdot 10^{32} \times 10^{-32} = 3.3.$$

Other experiments, in addition to Kamiokande, searched for signal events compatible with the proton decay (see Supplement 5.1). No plausible candidate was found, and a lower limit on the proton lifetime was set to be larger than 10^{33} years (at least in the decay channel $p \rightarrow e^+\pi^0$, in which it was thought to be easier to observe the proton decay).

Problem 13.3 The Earth mass is $M_T = 6 \cdot 10^{24}$ kg. The number N_p of protons (assumed to be 50% of the nucleons) is:

$$N_p = 0.5 \times \frac{M_T}{m_N} = 1.8 \cdot 10^{51}$$

where $m_N = 1.67 \cdot 10^{-27}$ kg is the nucleon (p or n) mass.

Let us assume that the whole energy produced inside the Earth is due to the proton decay, which transforms its rest mass ($m_p = 938$ MeV) into heat. Since $1\ \text{eV} = 1.6 \times 10^{-19}$ J, one has $E_{m_p} = 938\ \text{MeV} \times 1.6 \cdot 10^{-13}\ \text{J/MeV} = 1.5 \cdot 10^{-10}$ J. If τ_p denotes the proton lifetime, one has:

$$40\ \text{TW} = \frac{N_p \times E_{m_p}}{\tau_p} \longrightarrow \tau_p = \frac{1.8 \cdot 10^{51} \times 1.5 \cdot 10^{-10}\ \text{J}}{40 \cdot 10^{12}\ \text{J/s}} = 6.75 \cdot 10^{27}\ \text{s}.$$

Since 1 year $= 3.15 \cdot 10^7$ s, this corresponds to $\tau_p = 2.1 \cdot 10^{20}$ y.

This indirect limit is much worse than any direct experimental limit (see Supplement 5.1 and previous Problem). However, the obtained value is much longer that the Universe age!

A power of 40 TW corresponds to an energy flux on the Earth surface ($R_T = 6300\ \text{km} = 6.3 \cdot 10^8\ \text{cm}$) of $\Phi = \frac{40 \times 10^{12}\ \text{W}}{4\pi (6.3 \cdot 10^8)^2} = 8 \cdot 10^{-6}\ \text{J cm}^{-2}\,\text{s}^{-1}$. This can be compared with the $0.14\ \text{J cm}^{-2}\,\text{s}^{-1}$ arriving on Earth from the Sun.

Problem 13.4

(a) For the formation of a pair of X particles in the c.m. system, the c.m. energy must be equal to at least twice the X particle mass. The threshold energy is therefore $\sqrt{s} = 2M_X$. In the laboratory system, the four-momentum of the cosmic ray proton is $(E, \mathbf{p})$ and that of the target nucleon at rest is $(m, 0)$. In this case, one can write:

$$s = (E + m, \mathbf{p})^2 = E^2 + m^2 + 2Em - p^2 = 2m^2 + 2Em$$

and ($m = 1$ GeV):

$$E = \frac{s - 2m^2}{2m} = \frac{(2M_X)^2 - 2m^2}{2m} \simeq 2\frac{M_X^2}{m} = 2\frac{100^2}{1}\ \text{GeV} = 20\ \text{TeV}.$$

The corresponding momentum of the incident particle is $p \simeq E$.

(b) In the laboratory system, the four-momentum in the direction of motion of each particle is $(E', p') = (M_X\gamma', M_X\gamma'\beta')$. Using the momentum conservation law, one has:

$$p = 2M_X\gamma'\beta' \longrightarrow \gamma'\beta' = \frac{E}{2M_X} = \frac{2 \times 10^4}{200} = 100$$

from which one can derive:

$$100^2 = (\gamma'\beta')^2 = \frac{\beta'^2}{1-\beta'^2} \longrightarrow \beta' = 0.9999; \quad \gamma' = 100.$$

The energy of the X particle in the laboratory frame is $E' = 100M_X = 10$ TeV.

(c) The delay of the particle X with respect to a particle that propagates with speed c is:

$$\Delta t = \frac{L}{v} - \frac{L}{c} = \frac{L}{c\frac{v}{c}} - \frac{L}{c} = \frac{L}{c}\left(\frac{1}{\beta'} - 1\right) = \frac{2 \cdot 10^3}{3 \cdot 10^8} \times 1 \cdot 10^{-4} = 0.7 \text{ ns}.$$

Problem 13.5

(a) Consider an homogeneous sphere of density ρ and radius r, whose mass is $m_1 = \frac{4}{3}\pi r^3 \rho$. A spherical mass shell of radius $r + dr$ with mass $dm_2 = 4\pi r^2 dr\rho$ surrounding m_1 interacts with a gravitational potential $dU = G_N \frac{m_1 dm_2}{r}$. The work done by the gravitation to construct a sphere of radius R and mass $M = \frac{4}{3}\pi R^3 \rho$ corresponds to the gravitational binding energy:

$$|U| = \int_{r=0}^{r=R} dU = G_N \rho^2 \frac{4}{3}\pi 4\pi \int_{r=0}^{r=R} r^4 dr = \frac{3}{5}\frac{G_N}{R} M^2.$$

(b) The potential energy can be derived using the gravitational constant $G_N = 6.67 \times 10^{-11}$ m^3 kg^{-1} s^{-2} and recalling that 1 J $= \frac{1}{1.6 \cdot 10^{-19}}$ eV $= \frac{1}{1.6 \cdot 10^{-13}}$ MeV.

1. One light-year corresponds to $3 \cdot 10^8$ [m/s] $\times\, 3.15 \cdot 10^7$ [s] $\simeq 9 \cdot 10^{15}$ m. One solar mass is equivalent to $M = 2 \cdot 10^{30}$ kg. Therefore, one has:

$$|U_{protostar}| = \frac{3}{5}\, 6.67 \cdot 10^{-11} \frac{(2 \cdot 10^{30})^2}{9 \cdot 10^{15}} = 1.7 \cdot 10^{34} \text{ J} \simeq 10^{47} \text{ MeV}.$$

2. For the same mass in the solar radius $R = 7 \cdot 10^8$ m, one has:

$$|U_{star}| = \frac{3}{5} 6.67 \cdot 10^{-11} \frac{(2 \cdot 10^{30})^2}{7 \cdot 10^8} = 2.2 \cdot 10^{41} \text{ J} \simeq 1.4 \cdot 10^{54} \text{ MeV}.$$

3. For a white dwarf of radius $R = 1000$ km, one has:

$$|U_{WD}| = \frac{3}{5} 6.67 \cdot 10^{-11} \frac{(2 \cdot 10^{30})^2}{10^6} = 2 \cdot 10^{44} \text{ J} \simeq 1.3 \cdot 10^{57} \text{ MeV}.$$

4. For a neutron star of radius $R = 10$ km, one has:

$$|U_{NS}| = \frac{3}{5} 6.67 \cdot 10^{-11} \frac{(2 \cdot 10^{30})^2}{10^4} = 2 \cdot 10^{46} \text{ J} \simeq 1.3 \cdot 10^{59} \text{ MeV}.$$

The gravitational potential (or *binding*) energy of an object held together by gravity alone (which is negative), is the amount of energy required to pull all the material apart, to infinity. It is also the amount of energy that is liberated (usually in the form of heat) during the accretion of such an object from material falling from infinity.

In a star, there is roughly the same binding energy as there is thermal energy (this can be shown using the *virial theorem*). In white dwarfs, the thermal energy is unimportant, instead, it is the degeneracy energy of the electrons which is expected to be comparable to the gravitational binding energy. In neutron stars, it corresponds to the degeneracy energy of the neutrons (see Supplement 14.2). Nuclear physics is necessary to explain how stars maintain equilibrium against gravity (Chap. 14). Particle physics [13H97] is necessary to explain how energy is released during a gravitational stellar collapse, when 99% of the binding energy of a neutron star is released as neutrinos (see next Problem).

Problem 13.6 The main observable process is the antineutrino interaction with a free proton (the hydrogen nucleus): $\overline{\nu}_e p \to n e^+$

(a) From the definition $\gamma = \frac{1}{\sqrt{1-(v^2/c^2)}}$ and $E = \gamma m_\nu$ (using natural units with $c = 1$), one has:

$$\frac{v}{c} = \left(1 - \frac{1}{\gamma^2}\right)^{1/2} \simeq 1 - \frac{m_\nu^2}{2E^2}. \tag{13.7}$$

For a massive neutrino, one can write:

$$\frac{v_{max}}{c} = 1 - \frac{m_\nu^2}{2E_{min}^2}; \qquad \frac{v_{min}}{c} = 1 - \frac{m_\nu^2}{2E_{max}^2}. \tag{13.8}$$

This corresponds to a difference in speed ($\Delta v = v_{max} - v_{min}$) depending on the particle energy:

$$\frac{\Delta v}{c} = \frac{m_\nu^2}{2}\left(\frac{1}{E_{min}^2} - \frac{1}{E_{max}^2}\right). \tag{13.9}$$

A relatively large value of m_ν would lead to a sufficiently long delay between the arrival time of high and low energy neutrinos. In particular, the experiments would detect events with decreasing energy as a function of arrival time (higher energy neutrinos would arrive earlier). This was not observed, and it was therefore possible to set an *upper limit* on the neutrino mass. Assuming that the arrival time delay of supernova neutrinos on Earth depends on the different propagation speeds, one can write:

$$c\frac{\Delta t}{D} = c\frac{t_{max} - t_{min}}{D} = \frac{c}{v_{min}} - \frac{c}{v_{max}} = \frac{c\Delta v}{v_{min}v_{max}} \simeq \frac{\Delta v}{c} \tag{13.10}$$

assuming $v \simeq v_{max} \simeq v_{min} \simeq c$.

Equating Eq. (13.9) with Eq. (13.10), one has:

$$\frac{c\Delta t}{D} = \frac{\Delta v}{c} = \frac{m_\nu^2}{2}\left(\frac{1}{E_{min}^2} - \frac{1}{E_{max}^2}\right). \tag{13.11}$$

It can be concluded that $m_\nu = 0$ corresponds to $\Delta t = 0$. Experimentally, it is found that $\Delta t < 3$ s. Using the information given in the problem and $D = 1.7 \cdot 10^5 \times 3.15 \cdot 10^7 c$, the upper limit on the neutrino mass m_ν is:

$$m_\nu \leq \sqrt{\frac{2c\Delta t}{D \cdot (\frac{1}{E_{min}^2} - \frac{1}{E_{max}^2})}} = \sqrt{\frac{2 \cdot 3 \cdot c}{5.4\ 10^{12} \cdot c \cdot (\frac{1}{5^5} - \frac{1}{20^2})}} \simeq 5.5 \cdot 10^{-6}\ \text{MeV} \tag{13.12}$$

that is, $m_\nu \leq 6$ eV.

(b) The total energy released by the supernova as a neutrino burst is $\sim N_{tot} \cdot \langle E \rangle$ where the neutrino average energy $\langle E \rangle$ is about 12 MeV. Denoting as Φ_ν (cm^{-2}), the neutrino flux on Earth, the total number of neutrinos at the source (assuming an isotropic emission) is: $N_{tot} = \Phi_\nu \cdot 4\pi D^2$, where D is the distance from Earth to the source. The measured flux Φ_ν is obtained by requiring that $\Phi_\nu \sigma N_{tar}$ where σ is the interaction cross-section and N_{tar} is the number of free protons on the target, be equal to the number of observed events, that is about 10.
If the neutrinos interact only with the free protons, i.e., the hydrogen nuclei of the water, the cross-section is given by Eq. (8.28):

$$\sigma \simeq 7 \cdot 10^{-44}\ \text{cm}^2 \langle E \rangle^2 \simeq 1 \cdot 10^{-45}\ \text{m}^2.$$

The number N_{tar} of free proton on target is:

$$N_{tar} \simeq 2\frac{1000\ \text{t}}{18/N_A} \simeq 7 \cdot 10^{31}\ \text{protons} \tag{13.13}$$

(note the difference from Eq. (13.6)). Therefore, the measured flux on Earth is:

$$\Phi_\nu \simeq 10/(\sigma N_{tar}) \simeq 1.5 \cdot 10^{14}\ \text{m}^{-2}.$$

Finally, the total number of neutrinos and the total energy emitted at the source are:

$$N_{tot} = \Phi_\nu 4\pi D^2 \simeq 5 \cdot 10^{57}$$

$$\text{Total energy} \simeq N_{tot}\langle E_\nu \rangle \simeq 6 \cdot 10^{58}\ \text{MeV}.$$

(c) The sun mass is $M_\odot = 2 \cdot 10^{30}$ kg. The gravitational binding energy released to form a neutron star of radius 10 km and mass $1.4 M_\odot$ is (see Problem 13.5):

$$U_G = -\frac{3 \times 6.67 \cdot 10^{-11} \times (2.8 \cdot 10^{30})^2}{5 \times 10^4} \simeq -3.1 \cdot 10^{46}\ \text{J} \simeq -1.9 \cdot 10^{59}\ \text{MeV}.$$

Detailed studies show that, in the process of gravitational stellar collapse, about 99% of the gravitational binding energy is released as neutrinos.

Problem 13.7 In term of quark composition, the Δ^+ resonance with $m_\Delta = 1232$ MeV has the same structure as that of the proton (uud). The minimum energy is required when the Δ^+ is produced at rest: $\sqrt{s} = m_\Delta$. If p_p, p_γ are the four-momenta of the proton and photon, the resonance is produced when:

$$s = (p_p + p_\gamma)^2 = m_\Delta^2 \longrightarrow m_p^2 + 2E_\gamma E_p - 2\mathbf{p}_p\mathbf{p}_\gamma = m_\Delta^2. \tag{13.14}$$

For a photon, $|\mathbf{p}_\gamma| = E_\gamma$ and at high energy $E_p \simeq p_p$. In this case, Eq. (13.14) becomes:

$$2E_p(E_\gamma - E_\gamma \cos\theta) = m_\Delta^2 - m_p^2.$$

The minimum value for E_p occurs when $\theta = \pi$ (the directions of the photon and of the proton are opposite). In this case:

$$E_p = \frac{m_\Delta^2 - m_p^2}{4E_\gamma} = 1.6 \cdot 10^{20} \text{ eV} \tag{13.15}$$

where the values $E_\gamma = 10^{-3}$ eV, $m_\Delta = 1232$ MeV, $m_p = 938$ MeV are inserted in the last member.

(b) The mean free path of a proton in the CMBR is:

$$\lambda = \frac{1}{\sigma_{p\gamma}\rho_\gamma} = \frac{1}{250 \cdot 10^{-30} \times 400} = 10^{25} \text{ cm} = 3 \text{ Mpc}.$$

In the final equation, the distance is expressed in parsec, unit familiar to astronomers. The conversion is 1 pc $= 3 \cdot 10^{18}$ cm. Note that this is a small distance in terms of the dimensions of the universe.

Problem 13.9 The detection surface of a 10″ PMT is $A_{pmt} \simeq \pi r^2 \simeq 0.05$ m^2 where $r = 25.4/2$ cm is the PMT radius. Such PMTs have the advantage to fit inside commercial pressure-resistant glass spheres (called optical module, OM) and have been chosen by the IceCube and ANTARES neutrino telescopes (Fig. 13.3). The overall efficiency of an OM is somewhat reduced with respect to that of the PMT, due to the presence of glass, glue between the glass and the PMT, and mu-metal cage for magnetic shield. The overall efficiency is $\epsilon_{om} \simeq 0.8\epsilon_{pmt} = 0.2$.

The effective volume of a PMT is $V_{pmt} = A_{pmt} \times \lambda_{abs} \simeq 0.05$ [m^2] $\times$ 50 [m] $=$ 2.5 m^3. One Cherenkov photon falling inside this effective volume can produce a photoelectron (p.e.) with a probability $\epsilon_{om} \simeq 0.2$.

Let us call N_{pmt} the number of optical sensors inside the instrumented volume. The rate R between the effective PMT volume of N_{pmt} and the instrumented volume (1 km^3) is:

$$R = \frac{V_{pmt} \times N_{pmt}}{10^9 \text{ [m]}^3} = 2.5 \times 10^{-9} N_{pmt} \tag{13.16}$$

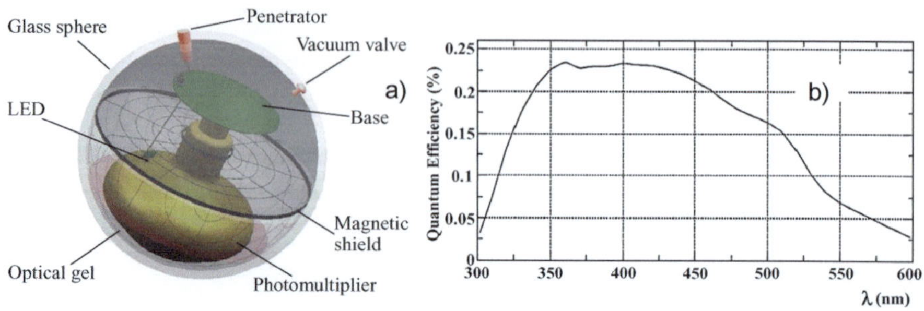

Fig. 13.3 (**a**) Sketch of an ANTARES optical module (OM). A large hemispherical (10 inches in diameter) photomultiplier (PMT) is protected by a pressure resistant glass sphere. The outer diameter of the sphere is 43.2 cm. A mu-metal cage protects the PMT from the Earth magnetic field. An internal LED is used for the calibration. (**b**) PMT quantum efficiency commonly used in ice or water (from Hamamatsu) [13C10]

Assuming that a muon emits 350 Cherenkov photons per cm of water, the total number of Cherenkov photons emitted by a muon with a 1 km long track, in the wavelength range of the PMT sensitivity, is $N_\gamma \simeq 3.5 \times 10^7$. The fraction converted into photoelectrons which gives a signal is:

$$\begin{aligned} N_{p.e.} &= N_\gamma \times R \times \epsilon_{om} \\ &\simeq \left(3.5 \times 10^7\right) \cdot \left(2.5 \times 10^{-9} N_{pmt}\right) \cdot \epsilon_{om} \\ &= 1.8 \times 10^{-2} N_{pmt} \end{aligned} \qquad (13.17)$$

We must take into account that in most cases many photons arrive on the same PMT during the integration window of the electronics (which is of the order of 20–50 ns). For this reason, an average of about $N_{p.e.} \sim 100$ p.e. are necessary to reconstruct the muon track. The number of optical sensors in a neutrino detector follows straightforward from Eq. (13.17):

$$N_{pmt} \simeq 100/1.8 \times 10^{-2} \simeq 5000 \qquad (13.18)$$

The IceCube collaboration has completed the construction of a telescope with 4800 OMs buried in ice; a neutrino telescope in the Mediterranean sea plans to use between 5000 and 10000 OMs, depending on the financial budget.

Problem 13.11 A fast rotating neutron star (if the rotation axis is different from the direction of the magnetic field) induces an electric field $\mathcal{E}$ through the Faraday law: $\nabla \times \mathcal{E} = \frac{1}{c}\frac{\partial B}{\partial t}$. From dimensional arguments, if L is the length of the region where the magnetic field changes,

$$\frac{\mathcal{E}}{L} = \frac{1}{c}\frac{dB}{dt},$$

the maximum energy E_{max} which an accelerated particle can gain is:

$$E_{max} = \int Ze\mathcal{E}dx = \int Ze\frac{L}{c}\frac{dB}{dt}dx = \int Ze\frac{L}{c}dB\frac{dx}{dt} = ZeLB\beta \qquad (13.19)$$

where $\beta = v/c$ is the particle velocity.

The condition dictated by Eq. (13.19) is due to Hillas, which derived E_{max} independently of the acceleration mechanism (Fig. 13.2).

The typical magnetic field of a star with radius $R = R_{\odot}$ is $B = 10^{-3} \div 10^{-2}$ T. After a gravitational collapse, the neutron star radius is of the order of $R_{NS} = 10$ km. By imposing conservation of the magnetic field flux, the magnetic field on the neutron star surface increases to $B_{NS} = (R/R_{NS})^2 B = 10^7 \div 10^8$ T.

By inserting numerical value in (13.19), for a relativistic ($\beta \simeq 1$) proton around a neutron star ($L = 10$ km), one obtains (c.g.s. units):

$$E_{max} = ZeLB\beta = 4.8 \cdot 10^{-10}\ [\text{esu}] \times 10^6\ [\text{cm}] \times 10^{11}\ [\text{Gauss}]$$
$$\simeq 5 \cdot 10^7\ \text{erg} \simeq 3 \cdot 10^{19}\ \text{eV}$$

where the conversion 1 eV $= 1.6 \cdot 10^{-12}$ erg is used in the last member.

The pulsar angular velocity is estimated in Supplement 14.2

Problem 13.12

(a) A muon with $E = 10$ GeV/c^2 energy has momentum $p = 10$ GeV/c (its mass can be neglected). The energy loss per unit path can be derived from Fig. 2.2(b): $(dE/dx)_\mu \simeq 2$ MeV g^{-1} cm^2. Note that we use the carbon curve as representative of a material made of light elements (see Problem 3.14). The liquid scintillator has density $\rho = 0.85$ g cm^{-3}. The energy loss of a through going particle crossing 25 cm of liquid scintillator is: $\Delta E_\mu \simeq (2 \cdot 25 \cdot 0.85) \simeq$ 43 MeV.

In 1931, Dirac introduced the magnetic monopole (MM) in order to explain the quantization of the electric charge, which follows from the existence of at least one free magnetic charge. He established the basic relationship between the elementary electric charge e and the basic magnetic charge g: $eg = n\hbar c/2$, where n is an integer, $n = 1, 2, \ldots$. The magnetic charge is $g = ng_D$, where $g_D = \hbar c/2e = 68.5e$ is the so-called unit Dirac charge. The specific energy loss for a MM with magnetic charge $g = g_D = 68.5e$ can be derived from Fig. 13.1(a) (lowest curve). At the speed $v_1 = 10^{-2}c$, one has $(dE/dx) \simeq 1$ [GeV/cm]/4.3 [g cm^{-3}] $\simeq 0.2$ GeV g^{-1} cm^2. The energy loss of a MM in the scintillator would be: $\Delta E \simeq (0.2 \cdot 25 \cdot 0.85)$ GeV $\simeq 4.3$ GeV, i.e., a factor about 100 times larger than that of a relativistic μ.

(b) From Fig. 13.1(b), for a muon, one has $(dL/dx)_\mu \simeq 0.05$ MeV/cm; it follows that $\Delta E_\mu = (0.05 \cdot 25)$ MeV $= 1.25$ MeV. The fraction of energy loss that produces visible light corresponds therefore to about $1.25/43 = 3\%$ for a particle near the ionization minimum, as it is the case for a relativistic muon. For a MM with $\beta = 10^{-2}$, one has $(dL/dx) \simeq 2$ MeV and $\Delta E_{MM} \simeq 50$ MeV, which corresponds to about $50/4300 = 1\%$ of the total energy loss. The processes that emit light largely increases for $\beta > 0.1$, as shown in Fig. 13.1(b).

(c) From Fig. 13.1(a), the energy loss for a MM with $\beta = 0.3$ is: $(dE/dx) \simeq$ 18 [GeV/cm]/4.3 [g cm^{-3}] $\simeq 4.2$ GeV g^{-1} cm^2. The corresponding energy loss in the scintillator would therefore be $\Delta E \simeq 4.1 \cdot 25 \cdot 0.85 \simeq 89$ GeV. If the MM behaves as a particle with an equivalent electric charge $e = g\beta$, one has:

$$(dE/dx)_{MM} \simeq (dE/dx)_\mu (g/e)^2 \beta^2$$

because the energy loss due to excitation-ionization depends on the square of the particle electric charge. This equation approximates the curve of Fig. 13.1(a) for $\beta \geq 0.1$. For $\beta = 0.3$, this would correspond to a muon with momentum:

$$p_\mu = m_\mu v\gamma = m_\mu \beta c\left(1-\beta^2\right)^{-1/2} \simeq 105 \cdot 0.3(1-0.09)^{-1/2} \simeq 33 \text{ MeV/c}.$$

Such a muon looses 7 MeV g^{-1} cm^2 (see Fig. 2.2(b)). Therefore, the corresponding energy loss is:

$$(dE/dx)_{MM} \simeq 0.007 \cdot (68.5)^2 \cdot 0.09 \simeq 3.0 \text{ GeV g}^{-1}\text{ cm}^2$$

which is not too different from the value of 4.2 GeV g^{-1} cm^2 obtained above.

References

[13H97] Herant, M., Colgate, S.A., Benz, W., Fryer, C.: Neutrinos supernovae. Los Alamos Science Number 25 (1997). http://la-science.lanl.gov/lascience25.shtml

[13O02] Ostrowski, M.: Mechanisms and sites of ultra high energy cosmic ray origin. Astropart. Phys. **18**, 229–236 (2002)

[13A08] Aharonian, F., et al.: High energy astrophysics with ground-based gamma ray detectors. Rep. Prog. Phys. **71**, 096901 (2008)

[13D08] de Angelis, A., et al.: Very-high-energy gamma astrophysics. Riv. Nuovo Cimento **31**, 187 (2008)

[13F09] Abdo, A., et al. (Fermi LAT Collaboration): Fermi Large Area Telescope Bright gamma-rays source list. Astron. Astrophys. Suppl. Ser. **183**, 46–66 (2009). arXiv:0902.1340 [astro-ph.HE]

[13C10] Chiarusi, T., Spurio, M.: High-energy astrophysics with neutrino telescopes. Eur. Phys. J. C **65**, 649–710 (2010). arXiv:0906.2634 [astro-ph.HE]

[13w3] http://icecube.wisc.edu

[13w4] http://www.km3net.org

Chapter 14
Fundamental Aspects of Nucleon Interactions

Problems

14.1. **Half-life and lifetime**. Derive the value of the half-life $t_{1/2}$ of a radioactive element when its lifetime τ is known.
[A. $t_{1/2} = \tau \ln 2$]

14.2. **Hydrogen isotopes**. The natural hydrogen is a mixture of two stable isotopes, hydrogen and deuterium. The deuterium nucleus has a binding energy of 2.23 MeV. The atomic mass of the natural hydrogen is 940.19 MeV. Calculate the relative abundance of the two isotopes in the natural hydrogen.
[See solutions]

14.3. **Nuclear fusion**. The deuterium nucleus ${}^2_1\mathrm{H}$ has a binding energy of 2.23 MeV. The tritium nucleus ${}^3_1\mathrm{H}$ has a binding energy of 8.48 MeV. Calculate the energy necessary to bring two deuterium nuclei to the distance of $r = 1.4 \cdot 10^{-13}$ cm. Estimate the corresponding temperature. These conditions are necessary to activate the fusion reaction:

$${}^2_1\mathrm{H} + {}^2_1\mathrm{H} \to {}^3_1\mathrm{H} + X$$

Indicate which particle X is produced in the final state and calculate the energy released in the fusion reaction.
[See solutions]

14.4. **Probability in radioactive decays**. The probability of a radioactive atom to decay in 1 second is equal to 5×10^{-11}. What is the probability that 5 decays take place in 1 second in a statistical sample made of 9×10^{10} atoms? And what is the probability for 15 decays?
[See solutions]

S. Braibant et al., *Particles and Fundamental Interactions: Supplements, Problems and Solutions*, Undergraduate Lecture Notes in Physics,
DOI 10.1007/978-94-007-4135-5_14, © Springer Science+Business Media Dordrecht 2012

14.5. **Radioactivity**. The α-decay of the radium isotope ($^{226}_{88}$Ra) has a half-life $t_{1/2} = 1602$ years. The unit of activity (1 Curie = 1 Ci) is defined as the number of disintegrations per second of a gram of radium.
(a) Write the decay reaction.
(b) Calculate the number of nuclei in one gram of $^{226}_{88}$Ra.
(c) Calculate the number of disintegrations per second corresponding to the activity of 1 Ci.
[See solutions]

14.6. **Carbon isotopes**. The natural carbon contains 98.89% of $^{12}_{6}$C and 1.11% of $^{13}_{6}$C, respectively having atomic masses $M(^{12}_{6}\text{C}) = 12.000$ u and $M(^{13}_{6}\text{C}) = 13.003$ u.
(a) Calculate the atomic mass of the natural carbon.
(b) A living organism contains a small fraction r of $^{14}_{6}$C with respect to the natural carbon. This fraction is $r = 1.3 \cdot 10^{-12}$. The $^{14}_{6}$C is a radioactive isotope which undergoes β^- decays with a half-life $t_{1/2} = 5730$ years. Calculate the activity of a gram of carbon in a living organism.
(c) The activity of a fossil containing a mass of carbon equal to $m = 5.000 \pm 0.005$ g is measured. If $n = 3600$ decays are recorded in 2 hours, calculate the age of the fossil. Also estimate the error on the measurement.
[See solutions]

14.7. **Nuclear reactor**. A nuclear reactor produces 2×10^9 Watts.
(a) Calculate the number of fission events occurring per second assuming

$$^{235}\text{U} \rightarrow {}^{A1}_{Z}\text{N} + {}^{A2}_{92-Z}\text{N} + neutrons + 200 \text{ MeV}.$$

(b) Estimate how many kilograms of uranium are consumed in one year knowing that the proton mass is $m_p = 938$ MeV and that the ^{235}U binding energy is ~8 MeV/nucleon.
[See solutions]

14.8. **Fermi momentum**. Calculate the Fermi momentum p_F and energy E_F of nucleons in the $^{16}_{8}$O nucleus. Assume a spherical nucleus with radius $R = 1.25A^{1/3}$ fm. The binding energy of the nucleus is 128 MeV. Calculate the depth of the potential well in the Fermi gas model. (Neglect the mass difference between proton and neutron).
[A. $p_F = 240$ MeV/c; $E_F = 30$ MeV; $U = 38$ MeV. See Sect. 14.3.1]

14.9. **Neutron moderation in nuclear reactors**. A nuclear reactor has a graphite moderator. The carbon nuclei can be effectively considered free to recoil when hit by fast neutrons. A fast neutron (1 MeV of kinetic energy) collides elastically against a nucleus of carbon 12.

(a) What are the initial speeds of the two particles in the center-of-mass system?
(b) In the c.m. system, the direction of the velocity of the carbon nucleus changes by 135° after the collision. What are the direction and kinetic energy of the neutron in the laboratory system after the collision?
(c) How many elastic collisions are needed on average for the neutron, assuming that the angular deviations are uniformly distributed in the c.m. system, so that its energy in the laboratory system is reduced from 1 MeV to 1 keV? Assume an average energy loss as the mean between the minimum and maximum values.

14.10. **Nuclear radioactive chain**. In a nuclear radioactive chain, τ_1 is the lifetime of the parent nucleus (type 1) in the decay $nucleus_1 \rightarrow nucleus_2$. The daughter nucleus (type 2) subsequently decays with a lifetime τ_2. Assuming that at the initial time ($t = 0$), the daughter nuclei are absent ($N_2(0) = 0$) and $N_1(0) = N_0$, determine the condition and the time necessary in order that the two nuclear activities become equal.
[See solutions]

14.11. **Nuclear binding energy-I**. Using the Weizsacker formula (14.15) for the nuclear binding energy:

$$\mathrm{BE} = a_0 A - a_1 A^{2/3} - a_2 \frac{Z^2}{A^{1/3}} - a_3 \frac{(A-2Z)^2}{A} \pm \frac{a_4}{A^{1/2}} \tag{14.1}$$

calculate the binding energy of isobar nuclei with $A = 27$: ${}^{27}_{12}\mathrm{Mg}$, ${}^{27}_{13}\mathrm{Al}$, ${}^{27}_{14}\mathrm{Si}$. Determine which is the more stable nucleus and indicate which terms of the binding energy formula make the other isotopes less stable.
[See solutions]

14.12. **Nuclear binding energy-II**. Using the Weizsacker formula (14.1) for the nuclear binding energy:
(a) Show that the ${}^{64}_{29}\mathrm{Cu}$ nucleus can decay either through β^+ and β^-; write the decay reactions.
(b) Calculate the maximum energy for the positron and the electron in each reaction.
(c) Which decay occurs with the largest probability?
[See solutions]

14.13. **Geiger-Nuttal law**. The isotopes of thorium ${}_{90}\mathrm{Th}$ decay by α emission ($m_\alpha = 3727.38$ MeV) to radium ${}_{88}\mathrm{Ra}$ isotopes. The measured lifetimes of three of these decays are reported in the last column of the following table. The binding energy (BE), the total angular momentum J and parity P of the nucleus before and after the decay are also indicated.

Z	A	BE (MeV)	J^P	Z	A	BE (MeV)	J^P	τ (s)
90	230	1755.22	0^+	88	226	1731.69	0^+	$3.4 \cdot 10^{12}$
90	229	1748.43	$5/2^+$	88	225	1725.30	$3/2^+$	$3.3 \cdot 10^{11}$
90	228	1743.19	0^+	88	224	1720.41	0^+	$8.7 \cdot 10^7$

Calculate the kinetic energy, momentum and angular momentum of the emitted α particles in the above thorium decays. Compare the obtained values with the energies and lifetimes in the Geiger-Nuttal plot, Fig. 14.9. One of the decays has lifetime about two orders of magnitude larger than the extrapolation of other data. Explain qualitatively why.
[See solutions]

14.14. **Measurement of geo-neutrinos**. Geophysical measurements show that the Earth emits approximately 40 TeraWatt of energy. Models predict that approximately 40% of this energy outflow is due to the decay of radioactive nuclei, 90% of which being due to the uranium and thorium decay chains. (A ^{238}U nucleus induces a cascade of 8 α and 6 β transitions and the chain ends in the stable ^{206}Pb isotope. The ^{232}Th induces a cascade of 6 α and 4 β transitions that terminates in the ^{208}Pb isotope.)

(a) Evaluate the flux on Earth surface of neutrinos emitted in the β decays of the U and Th chains (geo-neutrinos), assuming that neutrinos (with average energy of ~1 MeV) carry 10% of the uranium/thorium released energy.
(b) Guess which reaction and detector are needed to detect geo-neutrinos.

[See solutions]

14.15. **Nuclear muon capture**. Explain and discuss the process of nuclear capture of a μ^- in a hydrogen nucleus $\mu^- p \to n\nu_\mu$, and in a heavier nucleus, for example aluminum ($Z = 13$, $A = 27$). Consider that the lifetime is $\tau = 2.16$ μs for a free μ^-, while it is $\tau_{Al} = 0.88$ μs in aluminum.
Determine the μ^- mean free path in Al.
[See solutions]

14.16. **μ-mesic atom radius**.

(a) In the framework of Bohr's atomic theory, determine the radius R_0 of a mesic atom consisting of a proton and a μ^-.
(b) For a mesic atom consisting of a nucleus and a μ^-, calculate the atomic number Z for which the nuclear radius equate the size of the mesic atom (use Fig. 14.2 for the relationship between A and Z).

[A. (a) $R_0 = 256$ fm; (b) $Z \simeq 50$]

Supplement 14.1: Nuclear Collisions of Cosmic Rays During Propagation in the Galaxy

The astrophysical mechanisms that produce planetary systems (such as the solar system) and those that accelerate cosmic rays (CR) seem to be closely connected, as shown in Fig. 14.12. Both are related to the gravitational collapse of stellar objects with mass larger than our Sun ($M \sim 10M_{\odot}$). However, Fig. 14.12 shows a clear difference between the relative abundances of the Li, Be, B elements in the CRs with respect to that in the solar system. In particular, the ratio R between the number of Li, Be, B with respect to the C, N, O nuclei is $R_{CR} \simeq 0.25$ in the CRs, while it is $R_{ss} \sim 10^{-5}$ in the solar system. Li, Be, B act as catalysts of thermonuclear reactions in stars (Sect. 14.10.1) and a low abundance in the case of stellar collapse is expected, as the collapse occurs when material for fusion in the star is no longer available.

The presence of a relatively large number of Li, Be, B elements in the primary CRs can be explained by the fact that CRs propagate in the Galaxy before reaching the Earth. The interstellar medium is filled with matter (mainly hydrogen) and Li, Be, B elements are produced during propagation of heavier CR nuclei interacting with protons. To simplify the problem, let us globally identify the Li, Be, B elements with the symbol $\mathcal{L}$ (which stands for *light elements*) and the C, N, O elements with $\mathcal{M}$, *medium elements*.

The abundant $\mathcal{M}$-type CR nuclei propagate in the Galaxy, constantly deflected by galactic magnetic fields. Along the way, they interact with protons of the interstellar medium. This give rise to the so-called *spallation process*; the result of this interaction process is the expulsion of some nucleons from the hit nucleus. The spallation of $\mathcal{M}$ nuclei on protons produce $\mathcal{L}$ nuclei and can be quantitatively studied with accelerator data. The table below (from [14S90]) reports the spallation cross-section (in mb) for the reaction $p + X \rightarrow Y + anything$. The C, N, O elements are the target nuclei X and the Li, Be, B isotopes are the fragment nuclei Y. For instance, the probability that a proton interacting on a Carbon nucleus produces a $^{11}_{5}$B is $31.5/252.4 = 12.5\%$.

Nuclear			Target (X)		
Fragment (Y)			Fragmentation cross-section (mb)		
	Z	A	C	N	O
Li	3	6	12.6	12.6	12.6
Li	3	7	11.4	11.4	11.4
Be	4	7	9.7	9.7	9.7
Be	4	9	4.3	4.3	4.3
Be	4	10	2.9	1.9	1.9
B	5	10	17.3	16.0	8.3
B	5	11	31.5	15.0	13.9
Total cross-section (mb)			252.4	280.9	308.8

The average probability $P_{\mathcal{M}L}$ that a medium element produces a light element is 28%. This value is obtained by summing all the partial cross-sections in the table, divided by $252.4+280.9+308.8$ mb. Thus, during propagation:

$$N_{\mathcal{M}}+p\rightarrow N_{\mathcal{L}}+X \quad \text{with } P_{\mathcal{M}L}=0.28. \tag{14.2}$$

The mean free path (in $g\,cm^{-2}$) (see Problem 7.10) is $\lambda = A_{medium}/(N_A\sigma)$. N_A is Avogadro's number and σ is the nuclear cross-section on protons. For nuclei in the interstellar medium, $A_{medium}=1$. The nuclear radius scales with the cubic root of the atomic number A, Eq. (14.7), and the spallation cross-section is $\sigma\simeq(\pi R_0^2)A^{2/3}\simeq 45A^{2/3}$ mb. The cross-section and the mean free path of $\mathcal{M}$ and $\mathcal{L}$ nuclei are:

$$\sigma_{\mathcal{M}}\simeq 280\text{ mb}\longrightarrow\lambda_{\mathcal{M}}\simeq 6.0\text{ g cm}^{-2} \tag{14.3}$$

$$\sigma_{\mathcal{L}}\simeq 200\text{ mb}\longrightarrow\lambda_{\mathcal{L}}\simeq 8.4\text{ g cm}^{-2} \tag{14.4}$$

The problem is to determine the quantity of crossed galactic medium $\xi_T=x_T\rho$ that CRs must cross to reproduce the observed R_{CR} ratio between Li, B, Bo and C, N, O nuclei. x_T is the travelled distance (in cm) and ρ is the interstellar density (in $g\,cm^{-3}$). We can write a system of differential equations for the number of $\mathcal{M}$ and $\mathcal{L}$ elements as a function of path length (in $g\,cm^{-2}$) $\xi=x\rho$. The equation that describes the reduction of the number of $\mathcal{M}$ nuclei during their journey is:

$$\frac{d}{d\xi}N_{\mathcal{M}}(\xi)=-\frac{N_{\mathcal{M}}(\xi)}{\lambda_{\mathcal{M}}} \tag{14.5}$$

From the quoted astrophysical considerations, we can assume that $\mathcal{L}$ nuclei in CRs are not produced at the sources (stellar collapses), but following the spallation of heavier $\mathcal{M}$ elements. Their number increases with increasing path length of $\mathcal{M}$ nuclei. The differential equation that describes the number of produced $\mathcal{L}$ nuclei as a function of path length ξ contains a positive *source* term and a negative *attenuation* term:

$$\frac{d}{d\xi}N_{\mathcal{L}}(\xi)=+\frac{P_{\mathcal{M}L}}{\lambda_{\mathcal{M}}}N_{\mathcal{M}}(\xi)-\frac{N_{\mathcal{L}}(\xi)}{\lambda_{\mathcal{L}}} \tag{14.6}$$

The source term increases with probability $P_{\mathcal{M}L}$ as the spallation of $\mathcal{M}$ nuclei occur during propagation. The attenuation term is similar to that affecting the $\mathcal{M}$ nuclei in Eq. (14.5).

Equations (14.5) and (14.6) are coupled, since the number of $\mathcal{L}$ nuclei depends on $N_{\mathcal{M}}(\xi)$. Equation (14.5) can be immediately solved as:

$$N_{\mathcal{M}}(\xi)=N^0_{\mathcal{M}}e^{-\xi/\lambda_{\mathcal{M}}} \tag{14.7}$$

Some algebra is needed to solve (14.6). First, substitute $N_{\mathcal{M}}(\xi)$ with Eq. (14.7); then multiply both sides by $e^{\xi/\lambda_{\mathcal{L}}}$; the two terms containing $N_{\mathcal{L}}$ can be considered as the derivative of the product of two functions:

$$\frac{d}{d\xi}\left(N_{\mathcal{L}}(\xi)\cdot e^{\xi/\lambda_{\mathcal{L}}}\right)=\frac{P_{\mathcal{M}L}}{\lambda_{\mathcal{M}}}N^0_{\mathcal{M}}\cdot e^{(\xi/\lambda_{\mathcal{L}}-\xi/\lambda_{\mathcal{M}})} \tag{14.8}$$

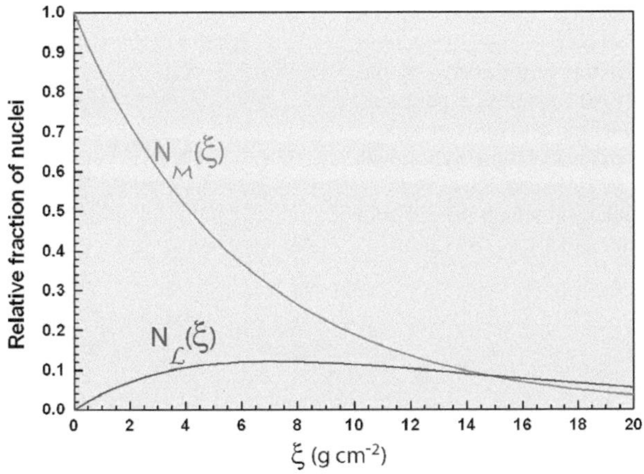

Fig. 14.1 Evolution of the number of $\mathcal{M}$ and $\mathcal{L}$ nuclei as a function of the path length ξ. Near the sources ($\xi = 0$) the $\mathcal{L}$ nuclei are absent. As ξ increases, $N_{\mathcal{L}}$ increases as light nuclei are produced by fragmentation of $\mathcal{M}$ nuclei. For instance, if the path length is equal to $\xi = 15\ \mathrm{g\,cm^{-2}}$, the ratio is $N_{\mathcal{L}}/N_{\mathcal{M}} = 1$. The measured ratio is $N_{\mathcal{L}}/N_{\mathcal{M}} \simeq 1/4$, corresponding to $\xi_T = x_T\rho \simeq 5\ \mathrm{g\,cm^{-2}}$

As the equation contains exponential functions, the *ansatz* is of the form $N_{\mathcal{L}}(\xi) = c \cdot (e^{-\xi/\lambda_{\mathcal{L}}} - e^{-\xi/\lambda_{\mathcal{M}}})$ where c is a constant to be determined with the boundary condition $N_{\mathcal{L}}(0) = 0$. By placing the test solution in Eq. (14.8), we obtain an identity if the constant c is $c = \frac{P_{\mathcal{M}L} \cdot N^0_{\mathcal{M}}}{\lambda_{\mathcal{M}}} \cdot \frac{\lambda_{\mathcal{M}}\lambda_{\mathcal{L}}}{\lambda_{\mathcal{L}} - \lambda_{\mathcal{M}}}$. The solution of (14.6) is:

$$N_{\mathcal{L}}(\xi) = \frac{P_{\mathcal{M}L}}{\lambda_{\mathcal{M}}} \cdot N^0_{\mathcal{M}} \cdot \frac{\lambda_{\mathcal{M}}\lambda_{\mathcal{L}}}{\lambda_{\mathcal{L}} - \lambda_{\mathcal{M}}} \cdot \left(e^{-\xi/\lambda_{\mathcal{L}}} - e^{-\xi/\lambda_{\mathcal{M}}}\right). \tag{14.9}$$

The two functions (14.7) and (14.9) are shown in Fig. 14.1, where the unknown parameter is assumed $N^0_{\mathcal{M}} = 1$.

The $N^0_{\mathcal{M}}$ parameter is not measurable (it represents how many $\mathcal{M}$ nuclei are emitted per second in the Galaxy from the sources). The measured quantity is the ratio of $R_{CR} = N_{\mathcal{L}}/N_{\mathcal{M}} = 0.25$, which does not depend on $N^0_{\mathcal{M}}$. The value of $\xi = \xi_T$ which gives the measured value of R_{CR} is determined using the ratio between Eqs. (14.7) and (14.9) (or through the inspection of Fig. 14.1):

$$\xi_T = x_T\rho = 5\ \mathrm{g\,cm^{-2}}. \tag{14.10}$$

This quantity is called the *average escape length* of CRs from our Galaxy. Since the value of the density of the interstellar material is $\rho \sim 0.3 \div 1\ \mathrm{cm^{-3}} = (0.3 \div 1) \times 1.6 \cdot 10^{-24}\ \mathrm{g\,cm^{-3}}$ [14K01], the travelled distance x corresponds to:

$$x_T = \xi_T/\rho = \frac{5\ \mathrm{g\,cm^{-2}}}{(0.3 \div 1) \times 1.6 \cdot 10^{-24}\mathrm{g\,cm^{-3}}} = (3 \div 10) \cdot 10^{24}\ \mathrm{cm} = 1 \div 3\ \mathrm{Mpc} \tag{14.11}$$

(1 parsec $= 3 \times 10^{18}$ cm).

In Supplement 1.1, the diffusion (or confinement) time τ_D of CRs in our Galaxy was introduced. From Eq. (14.11), the average value of τ_D can be derived:

$$\tau_D = x_T/c = \frac{(3 \div 10) \cdot 10^{24}\ \mathrm{cm}}{3 \cdot 10^{10}\ \mathrm{cm/s}} \simeq (1 \div 3) \times 10^{14}\ \mathrm{s} = 3 \div 10\ \mathrm{My}. \tag{14.12}$$

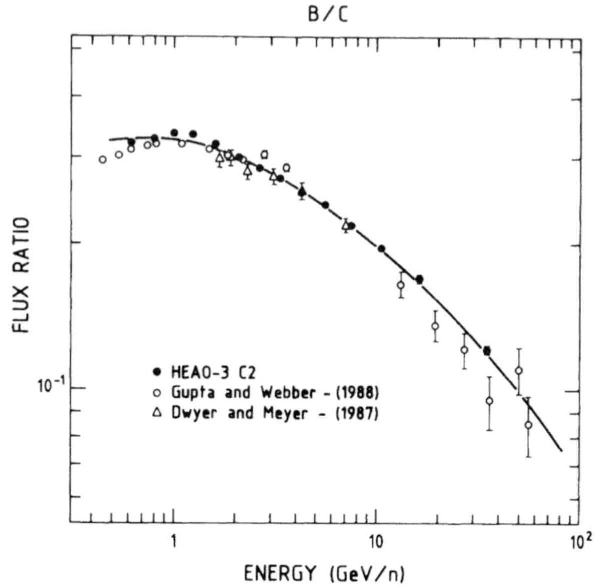

Fig. 14.2 Observed Boron to Carbon ratio, measured as a function of the energy by the HEA0-3 satellite experiment [14E90]. The *line* represent the result of a prediction with a *leaky box* model of our Galaxy in which the diffusion time of CRs is $\tau_D \sim E^{-0.6}$

As ξ_T depends only on R_{CR}, it does not depend on the observer's position: in any other position in the Galaxy, a hypothetical observer would measure the same $N_{\mathcal{L}}/N_{\mathcal{M}}$ ratio, obtaining the same value of τ_D.

The obtained value of x_T (14.11) is orders of magnitude larger than the thickness of the galactic disk (~300 pc). This confirms that the CRs are confined inside the Galaxy because they are constantly deflected by galactic magnetic fields. As the gyromagnetic radius for a particle with charge Ze, energy E, in the magnetic field B is $R \simeq \frac{E}{eZB}$, it is expected that the confinement time τ_D is not constant but decreases as the particle energy increases (at a fixed value of Ze). During propagation, higher energy particles have a larger probability to escape from the Galaxy due to their larger gyromagnetic radii (*leaky box* model of the Galaxy).

The energy-dependence of the diffusion time was derived by observing that the ratio between $\mathcal{L}$-to-$\mathcal{M}$ elements changes as a function of their energy E. Figure 14.2 shows the boron-to-carbon ratio (representing the $\mathcal{L}$ and $\mathcal{M}$ elements) as a function of E. As the energy increases, the escape probability $P_D \sim 1/\tau_D$ of C increases, and the number of secondary boron produced by C interaction on interstellar matter decreases. The dependence of τ_D on E is derived by adapting Monte Carlo simulations with the measured data, and cannot be easily analytically derived. In [14E90], it was found that the dependence on energy of the escape length (14.10) is:

$$\xi_T(E) = 34\left(\frac{E}{Z}\right)^{-0.6} \text{g cm}^{-2}. \tag{14.13}$$

This energy dependence was used in Supplement 1.1 to derive the energy spectrum of CRs in the proximity of the sources.

Supplement 14.2: Quantum Mechanics and Nuclear Physics → White Dwarfs and Neutron Stars

White Dwarfs When stars like our own sun die, they become *white dwarfs*, as $\sim 97\%$ of stars in our Galaxy which are not massive enough to become a neutron star. A white dwarf is a small star made of electron-degenerate matter. The material in a white dwarf no longer undergoes fusion reactions (Sect. 14.10), so the star has no source of energy, nor it is supported by the heat generated by fusion against gravitational collapse. It is supported only by electron degeneracy pressure, causing it to be extremely dense.

The Pauli exclusion principle disallows fermions from occupying the same quantum state. If one has a potential well (such as the one that in first approximation holds particles to form the star), fermions start filling up the quantum levels. The highest energy filled by the nuclei or electrons is called the Fermi energy (Sect. 14.3.1). Degenerate states occur when the Fermi energy is larger than the typical thermal energy.

Let consider first the case in which the degeneracy in a star is due to the atomic electrons. The degeneracy energy can be estimated using Heisenberg's uncertainty principle

$$\Delta x \cdot \Delta p \sim d \cdot p \sim \hbar.$$

If the number density is n [cm^{-3}], then each fermion is essentially confined to a cube of $d^3 \sim 1/n$. This implies that the typical momentum of an electron is $p \sim \hbar/d \sim \hbar n^{1/3}$. The energy of a typical electron is therefore:

$$E \sim \frac{p^2}{2m_e} \sim \frac{\hbar^2 n^{2/3}}{2m_e} \quad \text{in the non-relativistic case} \tag{14.14a}$$

$$E \sim pc \sim \hbar n^{1/3} c \quad \text{in the relativistic case.} \tag{14.14b}$$

If the collapsing star has N_e electrons, the total degeneracy energy of the star is $E_{deg} = E N_e$.

Let consider only the relativistic limit (14.14b). This limit is reached when the relative distance $\Delta x = d$ between electrons decreases and the energy increases up to the value $pc = m_e c^2$. Thus, from Heisenberg's principle, one can write:

$$\begin{aligned} & d\,(\mathrm{pc}) \sim \hbar c \rightarrow d \sim \frac{\hbar}{m_e c} \\ & \rho_{ec} = \frac{\mu}{d^3} = \frac{\mu}{(\hbar/m_e c)^3} \sim 3 \cdot 10^{13}\ \mathrm{g\,cm^{-3}} \end{aligned} \tag{14.15}$$

where ρ_{ec} is the *critical density of matter* for degenerate atomic electrons, and $\mu = m_p + m_e$. When the density ρ of the collapsing star is below ρ_{ec}, we are below the relativistic limit, and Eq. (14.14a) holds.

The total degeneracy energy of the star in the relativistic case (when the density reached ρ_{ec}) is obtained using Eq. (14.14b):

$$E_{deg} = EN_e = \hbar n^{1/3} c N_e \sim \frac{\hbar c M_*^{4/3} \eta^{4/3}}{R_* m_p^{4/3}} \tag{14.16}$$

where we have used the fact that the total number of electrons in a star with mass M_* and radius R_* is $N_e = M_* Z/A\mu = M_*\eta/\mu$ (where A is the atomic mass, Z is the number of electrons per atom and $\eta = Z/A$) and $n = N_e/V \sim N_e/R_*^3$.

Finally, we can use energy equipartition to estimate the equilibrium (see also Problem 13.5). This is done assuming that the gravitational binding energy $|U|$ is of order the degeneracy energy. Hence:

$$E_{deg} \sim |U| \longrightarrow \frac{\hbar c M_*^{4/3} \eta^{4/3}}{R_* \mu^{4/3}} \sim \frac{3}{5} \frac{G_N M_*^2}{R_*}. \tag{14.17}$$

Note that *the radius of the star cancels out!* As the mass M_* increases, the radius decreases and once white dwarfs become compact enough for the electrons to be relativistic, there is a solution with only one mass (which we indicate as the Chandrasekhar mass limit), irrespective of the radius.[1]

The final (universal) value of the mass from (14.17) is (neglecting the $3/5$ factor, as before, the $4\pi/3$ for the sphere volume was also neglected):

$$M_* \sim \left(\frac{\hbar c \eta^{4/3}}{G_N \mu^{4/3}}\right)^{3/2} = \left(\frac{\hbar c \eta^{4/3}}{G_N \mu^2}\right)^{3/2} \mu. \tag{14.18}$$

In the last equality, as M_* and μ have the same dimension (mass), the quantity between brackets is adimensional. If we consider hydrogen, $\mu \sim m_p$ and $\eta = 1$. The term $(\frac{\hbar c}{G_N m_p^2})$ appears in the textbook in Chap. 5, Eq. (5.2). It corresponds to the inverse of the gravitational coupling constant, α_G. Thus, Eq. (14.18) can be written as:

$$M_{Ch} \sim \frac{m_p}{\alpha_G^{3/2}} = \frac{1.6 \cdot 10^{-24}\ \text{g}}{(5.9 \cdot 10^{-39})^{3/2}} = 3.5 \cdot 10^{33}\ \text{g} = 1.4 M_\odot. \tag{14.19}$$

Equation (14.18) shows that $\hbar$ appears in the Chandrasekhar mass limit M_{Ch}: **the Planck constant not only determines the interaction of elementary particles, but also the mass scale and the inner structure of stars**. The evolution and structure of cosmic objects is determined by a known physics law and by the values of fundamental constants.

[1] We leave the student to work out the radius-mass relation for the non-relativistic case. When the density in a white dwarf is below ρ_{ec}, as its mass increases, its radius becomes smaller and smaller, scaling as $M_*^{-1/3}$. As the white dwarf approaches the mass limit M_{Ch}, the electrons become relativistic, and the dependence on mass becomes sharper than $-1/3$ as $M_* \to M_{Ch}$.

The typical radius of a white dwarf can be worked out from the ratio between the mass M_{Ch} and its density ρ_{ec}. Note that in (14.15), the quantity $\hbar/m_e c$ corresponds to the electron Compton wavelength $\lambda\!\!\!^{-}_e = 3.8 \cdot 10^{-13}$ m (see Appendix A5), and $\rho_{ec} = m_p/\lambda\!\!\!^{-3}_e$. Thus, one can write:

$$R_{WD} \sim \left(\frac{M_{Ch}}{\rho_{ec}}\right)^{1/3} = \left(\frac{m_p \lambda\!\!\!^{-3}_e}{\alpha_G^{3/2} m_p}\right)^{1/3} = \frac{\lambda\!\!\!^{-}_e}{\alpha_G^{1/2}} = \frac{3.8 \cdot 10^{-13}}{(5.9 \cdot 10^{-39})^{1/2}} \simeq 5 \times 10^6 \text{ m}. \tag{14.20}$$

The radius of a white dwarf is a few thousand kilometers, and depends on the electron Compton wavelength.

Neutron Stars A *neutron star* is a type of stellar remnant that can result from the gravitational collapse of a massive star ($M > 8M_\odot$) during a supernova event. As the core of a massive star is compressed during a supernova, increases in the electron Fermi energy allow the reaction (with threshold 1.36 MeV):

$$e^- + p \to n + \nu_e.$$

The weak interacting neutrinos escape, the matter cools down, the density increases and nuclei in the center of the star become neutron-enriched. At some point, the nuclei break into their components and enough neutrons were created so that they become degenerate. The neutron degeneracy pressure immediately stops the collapse and an equilibrium state is established. The transition from collapse to equilibrium is very sudden, and the infalling material experiences a *bounce* against the degenerate core, which creates an outward-propagating shock wave (the *supernova*). The shock wave (which is the mechanism involved in the cosmic rays acceleration) is further boosted by the neutrino pressure from the core [13H97].

From the mathematical point of view, the description of the neutron star equilibrium is similar to the case of the white dwarf. The degenerate fermion is now the neutron, and the critical density is obtained from (14.15) with the exchange $m_e \to m_n \sim \mu$, where m_n is the neutron mass. Thus, neutrons become relativistic when the density reaches the value:

$$\rho_{nc} = \frac{m_n}{(\hbar/m_n c)^3} = \frac{m_n}{\lambda\!\!\!^{-3}_n} \sim 10^{17} \text{ g/cm}^{-3}. \tag{14.21}$$

The total degeneracy energy has the same value as that given in Eq. (14.16)! The only change is to replace the number of electrons $N_e = M_* \eta/\mu$ with the number of neutrons: $N_n = M_*/m_n$, which corresponds to $\eta = 1$. In this condition, the same equality (14.17) between the degeneracy energy and binding energy $|U|$ holds, and the mass limit for a neutron star assumes the same value of the Chandrasekhar mass limit (14.19). The upper limit for a neutron star mass is the same as that of a white dwarf. As $\rho_{nc} \sim 10^3 \rho_{ec}$ the neutron star radius is much smaller than R_{WD} and from (14.20):

$$R_{NS} \sim \left(\frac{M_{Ch}}{\rho_{nC}}\right)^{1/3} = \left(\frac{m_p \lambda\!\!\!^{-}{}_n^3}{\alpha_G^{3/2} m_p}\right)^{1/3} = \frac{\lambda\!\!\!^{-}{}_n}{\alpha_G^{1/2}} = \frac{2.1 \cdot 10^{-16}}{(5.9 \cdot 10^{-39})^{1/2}} = 3 \times 10^3 \text{ m}. \tag{14.22}$$

The radius of a neutron star, object with a mass $M_{NS} \sim 1.4 M_\odot$, is of the order of a few km.

A **pulsar** is a rotating neutron star that emits a beam of electromagnetic radiation. The radiation can only be observed when the beam of emission is pointing towards the Earth. The rotation period and thus the interval between observed pulses is very regular, and the periods of their pulses range from 1.4 milliseconds to 8.5 seconds. This rotation slows down over time as electromagnetic radiation is emitted.

The millisecond rotating period for young pulsars can be estimated using basic physics arguments. A star like our sun has a radius $R \sim 7 \cdot 10^5$ km and rotates at 1 revolution per 30 days, so that the angular velocity is $\omega \sim 2.5 \cdot 10^{-6}$ rad/s. After the collapse, the neutron star has a radius $R_{NS} \sim 10$ km. For conservation of the angular momentum, one can write:

$$MR^2\omega = MR_{NS}^2\omega_{NS}$$

$$\omega_{NS} = \left(\frac{R}{R_{NS}}\right)^2 \times \omega = \left(\frac{7 \cdot 10^5}{10}\right)^2 \times 2.5 \cdot 10^{-6} = 12500 \text{ rad/s} \tag{14.23}$$

so that $T_{NS} = \frac{2\pi}{\omega_{NS}} = 0.5 \cdot 10^{-3}$ s.

Solutions

Problem 14.2 Using the proton and neutron mass, $m_p = 938.27$ MeV and $m_n = 939.56$ MeV, the deuterium nucleus has a mass: $m_d = 938.27 + 939.56 - 2.23 = 1875.6$ MeV. Denoting with x the ${}^1_1$H abundance, one can write:

$$m_p x + m_d(1-x) = 940.19 \longrightarrow x = 0.9980.$$

Consequently, the deuterium abundance is $(1-x) = 0.20\%$.

Problem 14.3 At the distance r, the Coulomb energy is given by $E_C = e^2/r$, where e is the elementary charge $e = 4.803 \times 10^{-10}$ esu. Using the conversion factors, 1 erg $= 10^{-7}$ J and 1 eV $= 1.602 \times 10^{-19}$ J, one has:

$$E_C = e^2/r = \left(4.8 \cdot 10^{-10}\right)^2 / 1.4 \cdot 10^{-13} = 1.65 \cdot 10^{-6} \text{ erg} \simeq 1 \text{ MeV}.$$

Using the Boltzmann constant $k = 8.617 \times 10^{-5}$ eV K^{-1}, this energy corresponds to a temperature of $T = E_C/k = 1.2 \times 10^{10}$ K.

The particle X corresponds to a proton (${}^1_1$H) and the Q value of the reaction is $Q = 8.48 - 2 \cdot 2.23 = 4.02$ MeV.

Problem 14.4 The decay (or emission) probability in one second is the *activity* of the sample, $A = 5 \times 10^{-11}\ \text{s}^{-1}$. The average number of decays per second for a sample of $N_0 = 9 \times 10^{10}$ atoms is:

$$\lambda = N_0 \times A = 4.5\ \text{s}^{-1}.$$

This corresponds to the average value of a Poisson statistical distribution. The Poisson statistics gives the probability that a certain number n of events occurs when the average probability is λ:

$$P(n) = e^{-\lambda}\frac{\lambda^n}{n!}. \tag{14.24}$$

For $n = 5$, the probability is $P(5) = 17\%$, while for $n = 15$, it is $P(15) = 5 \times 10^{-5}$.

Problem 14.5

(a) The radium decays to radon:

$$^{226}_{88}\text{Ra} \rightarrow {}^{222}_{86}\text{Rn} + {}^{4}_{2}\text{He}.$$

(b) The number of nuclei per gram of material is:

$$N = N_A/226 = 2.66 \times 10^{21}$$

where $N_A = 6.022 \times 10^{23}\ \text{mol}^{-1}$ is Avogadro's number.

(c) The activity (14.21) of 1 g of radium ($1\ \text{y} = 3.15 \cdot 10^7$ s) is:

$$\begin{aligned} A &= N/\tau = N \ln 2/t_{1/2} \\ &= 2.66 \times 10^{21} \cdot 0.69/\left(1602 \cdot 3.15 \times 10^7\right) = 3.7 \times 10^{10}\ \text{s}^{-1}. \end{aligned}$$

Problem 14.6

(a) The natural carbon atomic mass is:

$$M(\text{C}) = (98.89 \times 12.000 + 1.11 \times 13.003)/100 = 12.011\ \text{u}.$$

(b) The $^{14}_{6}\text{C}$ lifetime is $\tau = t_{1/2}/\ln 2 \simeq 8270$ y. The number of $^{14}_{6}\text{C}$ isotopes present in one gram of carbon is: $N = 1.3 \times 10^{-12} N_A/A = 6.5 \times 10^{10}$. The carbon activity is:

$$A_0 = N/\tau = 6.5 \times 10^{10}/\left(8270\ [\text{y}] \cdot 3.15 \times 10^7\ [\text{s/y}]\right) = 0.25\ \text{s}^{-1}\,\text{g}^{-1}$$

(c) With 3600 disintegrations measured in 2 hours (7200 s) in a 5 g sample, the activity of one gram of the fossil carbon is $A(t^*) = 3600/(5 \cdot 7200) = 0.1\ \text{s}^{-1}\,\text{g}^{-1}$. The ratio between the activity of the fossil and of the new carbon sample is:

$$A(t^*)/A_0 = e^{-t^*/\tau} \rightarrow \ln\left[A(t^*)/A_0\right] = -t^*/\tau.$$

The age of the fossil is therefore:

$$t^* = -\tau \ln(0.4) = 7580 \text{ y}.$$

The main error comes from Poisson fluctuations of counts; the percentage error corresponds to $\Delta n/n = \sqrt{3600}/3600 = 1.67\%$, while $\Delta m/m \sim 0.1\%$. The corresponding error on the age determination is about 130 y.

Problem 14.7

(a) The power $P = 2$ GW in this problem corresponds to that of a large nuclear power plant. Each elementary reaction releases $Q = 200 \text{ MeV} = 200 \cdot 1.6 \times 10^{-13} \text{ J} = 3.2 \times 10^{-11}$ J. The number N of reactions per second required to maintain a power of 2 GW is then:

$$N = P/Q = \frac{2 \cdot 10^9 \text{ J/s}}{3.2 \cdot 10^{-11} \text{ J}} = 6.3 \cdot 10^{19} \text{ reactions/s}.$$

(b) The mass of a ^{235}U nucleus can be approximated as that of 235 protons ($m_p = 938$ MeV) minus $\sim$8 MeV/nucleon of binding energy:

$$\begin{aligned} m_U = 235 \cdot (938 - 8) &\simeq 2.2 \times 10^5 \left[\text{MeV/c}^2\right] \\ &\simeq 2.2 \times 10^{11} \left[\text{eV/c}^2\right] \\ &\simeq 3.9 \times 10^{-25} \text{ [kg]} \end{aligned}$$

using the conversion factor: $1 \text{ [eV/c}^2] = 1.6 \times 10^{-19} \text{ [J]}/(3 \times 10^8 \text{ [m/s]})^2 = 1.8 \cdot 10^{-36}$ [kg]. The fuel mass M_{1y} used in one year ($T = 3.15 \times 10^7$ s) is:

$$M_{1y} = N \cdot m_U \cdot T = 6.3 \times 10^{19} \cdot 3.9 \times 10^{-25} \cdot 3.15 \times 10^7 = 774 \text{ kg}.$$

Problem 14.10 The activities of the two nuclei are given in Eq. (14.25)

$$\begin{aligned} A_1(t) &= N_0 \lambda_1 e^{-\lambda_1 t} \\ A_2(t) &= N_0 \frac{\lambda_1 \lambda_2}{\lambda_2 - \lambda_1} \left(e^{-\lambda_1 t} - e^{-\lambda_2 t}\right) \end{aligned}$$

with $\lambda_1 = 1/\tau_1$; $\lambda_2 = 1/\tau_2$. The two activities are equal at the time $t = t^*$ such that $A_1(t^*) = A_2(t^*)$, that is:

$$N_0 \lambda_1 e^{-\lambda_1 t^*} = N_0 \frac{\lambda_1 \lambda_2}{\lambda_2 - \lambda_1} \left(e^{-\lambda_1 t^*} - e^{-\lambda_2 t^*}\right)$$

After some algebra, one finds:

$$\lambda_1/\lambda_2 = e^{-(\lambda_2 - \lambda_1) t^*}$$

and

$$t^* = \ln(\lambda_2/\lambda_1)/(\lambda_2 - \lambda_1). \tag{14.25}$$

This equation makes sense if $(\lambda_2 - \lambda_1) > 0$, or (equivalently) $\tau_2 < \tau_1$ (this is the situation depicted in Fig. 14.1, considering $\lambda_1 = \lambda_{\mathcal{M}}$; $\lambda_2 = \lambda_{\mathcal{L}}$). In this case, a type 2 nucleus decays faster than the parent nucleus; nucleus 2 activity, null at $t = 0$, becomes equal to that of the type 1 nucleus at a time t^* and then decreases. For $t \gg t^*$, equilibrium is reached when the ratio between the activities is approximately constant:

$$\frac{A_2(t)}{A_1(t)} \longrightarrow \frac{\lambda_2}{\lambda_2 - \lambda_1} \tag{14.26}$$

(this equation can be derived neglecting the term $e^{-\lambda_2 t}$ in $A_2(t)$ in the limit $t \longrightarrow \infty$). This is the *transient equilibrium* situation. If $\tau_2 \ll \tau_1$, type 2 nuclei decay immediately after the formation and activities are approximately equal $\lambda_2 N_2 = \lambda_1 N_1$. This situation is called *secular equilibrium.*

If $(\lambda_2 - \lambda_1) \leq 0$ (or $\tau_2 \geq \tau_1$), Eq. (14.25) has no real solutions. In this case, the activity of type 2 nuclei increases rapidly due to the type 1 nucleus decays and reaches the maximum value at time $t^* = (\ln \lambda_1/\lambda_2)/(\lambda_1 - \lambda_2)$. At time $t \gg t^*$, the number of type 1 nuclei is much reduced and the activity of type 2 nuclei decreases exponentially with lifetime τ_2. In this case, equilibrium between the activities is not reached.

Problem 14.11 The binding energy is obtained from Eq. (14.1) using the coefficients given in Table 14.2: ($a_0 = 15.74$, $a_1 = 17.61$, $a_2 = 0.71$, $a_3 = 23.42$) MeV. $a_4 = 0$ since the nucleon is odd ($A = 27$). The values of each term of Eq. (14.1) are (in MeV):

Z	$a_0 A$	$-a_1 A^{2/3}$	$-a_2 \frac{Z^2}{A^{1/3}}$	$-a_3 \frac{(A-2Z)^2}{A}$	Sum (MeV)
12	425.0	−158.1	−34.1	−7.8	224.9
13	425.0	−158.1	−40.0	−0.9	225.9
14	425.0	−158.1	−46.4	−0.9	219.5

The more stable nucleus is $^{27}_{13}$Al, with the largest binding energy. The first two terms (volume and surface terms) depending only on A are equal for the three nucleus. The Coulomb term decreases the binding energy as Z increases. The asymmetry term has the larger (absolute) value for the $^{27}_{12}$Mg, because the largest difference between the number of neutrons and protons, $|2Z - A| = 3$. This term is equal for $^{27}_{13}$Al and $^{27}_{14}$Si ($|2Z - A| = 1$).

Problem 14.12

(a) The two decays are: ${}^{64}_{29}\mathrm{Cu} \to {}^{64}_{28}\mathrm{Ni}\, e^+ \nu_e$ and ${}^{64}_{29}\mathrm{Cu} \to {}^{64}_{30}\mathrm{Zn}\, e^- \overline{\nu}_e$.
(b) The maximum energy is $T_{max} = 0.93$ MeV for the positron of the β^+ decay and $T_{max} = 0.67$ MeV for the electron of the β^- decay.
(c) The β^+ decay, because of the larger energy available in the final state.

Problem 14.13 The binding energy of the α particle is very high:

$$\mathrm{BE}(\alpha) = 2m_p + 2m_n - m_\alpha = 28.3 \text{ MeV}.$$

The energy E_α of the α particle (by energy conservation) is given by the difference between

$$E_\alpha = \mathrm{BE}(\alpha) - \left[\mathrm{BE(Th)} - \mathrm{BE(Ra)}\right]$$

for the three different isotopes. Inserting the numerical values given in the table, one obtains: $E_\alpha = 4.77$, 5.17, 5.52 MeV for thorium isotopes with $A = 230$, 229, 228, respectively.

For the two $0^+ \to 0^+$ transitions (thorium with $A = 230, 228$) no changes in the nuclear angular momentum are involved. The emitted α particle has no orbital angular momentum, and thus $T_\alpha = E_\alpha$, where T_α is the kinetic energy. In this case, the momentum of the alpha particle can be determined by the non-relativistic relation $p_\alpha = \sqrt{2T_\alpha m_\alpha}$. Consequently, $p_\alpha = 188$ MeV/c in the case of ${}^{230}_{90}\mathrm{Th}$ and $p_\alpha = 201$ MeV/c in the case of ${}^{228}_{90}\mathrm{Th}$.

For the ${}^{229}_{90}\mathrm{Th}$, the α particle must have a non-zero orbital angular momentum (for angular momentum conservation). Thus, the particle must have rotational energy, accordingly causing a reduction in the kinetic energy. Therefore, $T_\alpha < E_\alpha$, and only an upper limit on the linear momentum can be derived ($p_\alpha < 195$ MeV/c).

This fact explains the reason why the lifetime of the ${}^{229}_{90}\mathrm{Th}$ is much higher than expected ($\tau \sim 10^8$ s) using the value of $E_\alpha = 5.17$ MeV in Fig. 14.9: the quantity which is relevant for the tunneling effect is the kinetic energy of the α particle, not the total energy (which includes the rotational one).

Problem 14.14

(a) A power of 40 TW corresponds to an energy flux on the Earth surface ($R_T = 6300\text{ km} = 6.3 \times 10^8$ cm) of

$$\Phi = \frac{40 \times 10^{12}\text{ W}}{4\pi (6.3 \times 10^8)^2} = 8 \times 10^{-6}\text{ J cm}^{-2}\text{s}^{-1}.$$

Assuming that 40% of that energy is due to nuclear reactions, and that 10% is carried by $E_\nu = 1$ MeV neutrinos (1 J $= 6.25 \times 10^{12}$ MeV):

$$\Phi_\nu = 0.4 \cdot 0.1 \cdot \frac{\Phi \cdot 6.25 \times 10^{12}}{E_\nu} = 2 \times 10^6 \frac{\nu}{\text{cm}^2\text{ s}}.$$

(b) The geo-neutrino are electron antineutrinos ($\overline{\nu}_e$), as those emitted by a nuclear reactor. Antineutrinos above a certain energy threshold can be detected using the same technique as that used by Cowan and Reines for the first neutrino observation (see Sect. 8.5). Compared to the reactor neutrino flux at the Cowan and Reines experiment, the geo-neutrino flux is about a factor $\sim 10^7$ smaller. A large enough number of measured events can be obtained only by:

- using a larger detector;
- increasing the observation lifetime;
- reducing the energy threshold to detect ν_e.

All the above condition were fulfilled by the Borexino detector at Laboratori Nazionali del Gran Sasso (Italy) which observed geo-neutrinos through the inverse-beta reaction (8.19) with a 252.6 ton yr exposure. The low-energy $\overline{\nu}_e$ threshold of 1.8 MeV was allowed by the unprecedentedly low intrinsic radioactivity achieved in Borexino and by the high photon yield of the used liquid scintillator [14B10].

Problem 14.15 In analogy with the Bohr atomic theory, one can calculate the radius of a muonic atom (i.e., an atom in which an electron is replaced by a captured μ^-). For a nucleus with Z protons:

$$r_Z^{\mu} = \frac{r_0}{Z} \cdot \frac{m_e}{m_\mu}$$

where $r_0 \simeq 0.53$ Å is the Bohr radius which (Appendix A5) is inversely proportional to the mass of the orbiting particle. For a mesic atom, the radius of the orbit is a factor $m_e/m_\mu = 1/207$ smaller. For the ground state of the hydrogen atom, for example, one has: $r_0^{\mu} = 2.56 \cdot 10^{-3}$ Å.

Let us define τ_{free} the lifetime of a free muon in vacuum. When in matter the competitive process of the capture of negative muons by atomic nuclei is present, with lifetime τ_{capt}, the effective lifetime is reduced. Following Eq. (4.44), the effective muon lifetime τ corresponds to:

$$1/\tau = 1/\tau_{free} + 1/\tau_{capt}.$$

In other words, $1/\tau_{capt}$ represents the probability per time unit of the capture process, while $1/\tau_{free}$ is that of spontaneous decay. Since the measured lifetime of μ^- in aluminum is 0.88 μs, one obtains $\tau_{capt} \simeq 1.49$ μs.

The μ^- capture process ($\mu^- p \to n\nu_\mu$) in hydrogen or in a heavier nucleus depends on the fraction of time that the muon is "inside" the nuclear region (as the weak interaction process can be considered point-like). Let us assume that the captured muon orbits in the ground state, with spherical symmetry. The probability density of finding the muon is thus constant inside the sphere with radius r_Z^{μ}. The probability f that the muon is present inside the nucleus is proportional to the ratio between the volume of the nucleus and that of the orbital radius r_Z^{μ}, i.e.:

$f \simeq (V_{nucleus}/V_{orbit}) = (R_A/r_Z^{\mu})^3$, where R_A is the radius of a nucleus with atomic weight A. This radius is given in Eq. (14.7): $R_A \simeq 1.2 \times 10^{-5} \sqrt[3]{A}$ (in Å). The fraction of time that the muon spends "inside" the nuclear matter, where the muon can be captured by a proton, is:

$$f \simeq \left(R_A/r_Z^{\mu}\right)^3 \simeq 1.03 \times 10^{-7} AZ^3.$$

For the hydrogen ($Z = A = 1$), $f_H = 1.03 \times 10^{-7}$; for the aluminum ($Z = 13$, $A = 27$), $f_{Al} \simeq 0.61 \times 10^{-2}$. The muon capture probability in hydrogen is therefore relatively small.

One can now calculate the mean free path λ_{capt} of negative muons. It depends on the lifetime τ_{capt} of the capture process, on the muon speed v and on the fraction of time f in which the muon is inside the nucleus:

$$\lambda_{capt} = f \tau_{capt} v,$$

where $v = Zc\alpha = Zc/137$ is the muon speed in an atomic orbit (from the atomic theory, similarly to electrons). Numerically, the mean free path in aluminum is $\lambda_{capt} \simeq 26$ cm.

References

[14E90] Engelmann, J.J., et al.: Charge composition and energy spectra of cosmic-ray nuclei for elements from Be to Ni. Results from HEAO-3-C2. Astron. Astrophys. **233**, 96–111 (1990)

[14S90] Silberberg, R., Tsao, C.H.: Spallation processes and nuclear interaction products of cosmic rays. Phys. Rep. **191**, 351–408 (1990)

[14K01] Ferriere, K.: The interstellar environment of our galaxy. Rev. Mod. Phys. **73**, 1031–1066 (2001)

[14B10] Borexino Collaboration: Observation of geo-neutrinos. Phys. Lett. B **687**, 299–304 (2010). Also: arXiv:1003.0284v2 [hep-ex]

References[1]

General Bibliography

[L87] Leo, W.R.: Techniques for Nuclear and Particle Physics Experiments. Springer, Berlin (1994). ISBN: 978-3540572800

[P87] Perkins, D.H.: Introduction to High Energy Physics, 4th edn. Cambridge University Press, Cambridge (2000). ISBN: 978-0521621960

[A89] Aitchison, I.J.R., Hey, A.J.G.: Gauge Theories in Particle Physics, 3rd edn. Taylor & Francis, New York (2002). ISBN: 978-0750308649

[G90] Gaisser, T.K.: Cosmic Rays and Particle Physics. Cambridge University Press, Cambridge (1991). ISBN: 978-0521339315

[H91] Hughes, I.S.: Elementary Particles, 3rd edn. Cambridge University Press, Cambridge (1991). ISBN: 978-0521407397

[L92] Longair, M.S.: High Energy Astrophysics, 2nd edn. Cambridge University Press, Cambridge (1992). ISBN: 978-0521387736 (Vol. I); ISBN: 978-0521435840 (Vol. II)

[P95] Povh, B., Rith, K., Scholz, C., Zetsche, F.: Particles and Nuclei. An Introduction to the Physical Concepts, 6th edn. Springer, Berlin (2008). ISBN: 978-3540793670

[J99] Jackson, J.D.: Classical Electrodynamics. Wiley, New York (1999). ISBN: 978-0471309321

[C03] Ceradini, F.: Appunti del corso di Istituzioni di Fisica Nucleare e Subnucleare (in Italian). http://webusers.fis.uniroma3.it/~ceradini/efns.html

[1]This *Problems and solutions* book refers to *Particle and Fundamental Interactions* [B11], and the bibliography reported therein should also be considered. Below there is a list of texts, reviews and some specialized works which were explicitly considered in this book. The more general bibliography is referred to within square brackets and ordered by year (the last two characters). The first character is that of the first author surname. Those more specific, used for a particular Problem or Supplement, are divided between chapters, whose number is denoted by the first characters. We frequently refer to web pages: in this case, they are indicated by the lower case character [w], followed by a progressive number. The Nobel lectures are all available at: http://www.nobelprize.org/nobel_prizes/physics/laureates/.

S. Braibant et al., *Particles and Fundamental Interactions: Supplements, Problems and Solutions*, Undergraduate Lecture Notes in Physics,
DOI 10.1007/978-94-007-4135-5, © Springer Science+Business Media Dordrecht 2012

[S04] Sokolsky, P.: Introduction to Ultrahigh Energy Cosmic Ray Physics. Westview Press, Boulder (2004). ISBN: 978-0813342122

[P10] Nakamura, K., et al. (Particle Data Group): 2010 review of particle physics. J. Phys. G **37**, 075021 (2010). Physics Lett. B **667**, 1 (2008). Available at: http://pdg.lbl.gov/

[B11] Braibant, S., Giacomelli, G., Spurio, M.: Particle and Fundamental Interactions. Springer, Berlin (2011). ISBN: 978-9400724631

Index

S. Braibant et al., *Particles and Fundamental Interactions: Supplements, Problems and Solutions*, Undergraduate Lecture Notes in Physics,
DOI 10.1007/978-94-007-4135-5, © Springer Science+Business Media Dordrecht 2012

MIX
Papier aus verantwortungsvollen Quellen
Paper from responsible sources
FSC® C105338

If you have any concerns about our products,
you can contact us on
ProductSafety@springernature.com

In case Publisher is established outside the EU,
the EU authorized representative is:
Springer Nature Customer Service Center GmbH
Europaplatz 3, 69115 Heidelberg, Germany

Printed by Libri Plureos GmbH
in Hamburg, Germany